Newnes Radio Amateur and Listener's Data Handbook

Newnes Radio Amateur and Listener's Data Handbook

Steve Money

Newnes
An imprint of Butterworth-Heinemann Ltd
Linacre House, Jordan Hill, Oxford OX2 8DP

Ⓡ A member of the Reed Elsevier plc group

OXFORD   LONDON   BOSTON
MUNICH   NEW DELHI   SINGAPORE   SYDNEY
TOKYO   TORONTO   WELLINGTON

First published 1995

British Library Cataloguing in Publication Data
Money, Steve A.
  Newnes Radio Amateur and Listener's Data
  Handbook. – 2 Rev.ed
  I. Title
  621.384151

ISBN 0 7506 2094 3

Library of Congress Cataloguing in Publication Data
Money, Steve A.
  Newnes radio amateur and listener's data handbook/Steve Money.
  p.   cm.
  Includes bibliographical references and index.
  ISBN 0 7506 2094 3
  1. Radio – Amateurs' manuals. I. Title.
  TK9956.M64   94–23260
  621.3841'6–dc20   CIP

Composition by Scribe Design, Gillingham, Kent, UK
Printed in Great Britain by Clays Ltd, St Ives plc

Contents

1

Radio waves and signal definitions

All communication by radio depends upon the transmission of information via high-frequency electromagnetic waves. An electromagnetic wave consists of a high-frequency electric field and its associated magnetic field. Each field has a sinusoidal waveform, and the number of cycles occurring in one second is known as the frequency of the wave, which is expressed in hertz (Hz). Thus one hertz represents a frequency of one cycle per second.

Radio frequencies start at a few thousand hertz and may be more conveniently expressed in terms of kilohertz (kHz), where 1 kHz = 1000 Hz. For higher frequencies the unit of megahertz (MHz) is used, where 1 MHz = 1000 kHz. At the upper end of the radio frequency spectrum, frequencies are usually expressed in gigahertz (GHz), where 1 GHz = 1000 MHz. The electromagnetic frequency spectrum continues beyond the radio wave region and includes infrared light, visible light, ultraviolet light and eventually X-rays.

The alternating electric and magnetic fields produce a wavefront which travels outward from the source of the field at the speed of light; the distance travelled during one cycle of the electromagnetic field is known as the wavelength.

The amplitude of the field is usually expressed in terms of the amplitude of the electrical component in units of volts per metre. For some applications, such as electromagnetic compatibility, the magnetic component of the field is important and its amplitude is expressed in units of webers per square metre.

Frequency–wavelength conversion

Since the speed of light is a constant of approximately 300 000 000 metres per second, the frequency is related to wavelength by the equation:

$$\text{Frequency } f = \frac{300}{\lambda} \text{ MHz}$$

This equation can be rearranged to give a new equation for wavelength in terms of frequency as follows:

$$\text{Wavelength} = \frac{300}{f} \text{ metres}$$

The above equations are true for an electromagnetic wave travelling in free space. If the wave is travelling along a cable or through materials such as water, its velocity is reduced slightly, so that the wavelength becomes longer than it would be in free space for the same wave frequency.

Radio frequency band names

Radio waves can have frequencies ranging from a few kilohertz up to hundreds of gigahertz, where they merge into the region of infrared light. For convenience the frequency range is divided into a series of bands which are given individual names as follows:

3–30 kHz	Very low frequency (VLF)
30–300 kHz	Low frequency (LF)
300–3000 kHz	Medium frequency (MF)
3–30 MHz	High frequency (HF)
30–300 MHz	Very high frequency (VHF)
300–3000 MHz	Ultra high frequency (UHF)
3–30 GHz	Super high frequency (SHF)
Above 30 GHz	Extra high frequency (EHF)

Radio wavelength band names

An alternative way of naming the radio bands is in terms of their wavelength. In the following table the corresponding frequency band is shown in brackets.

10 000–100 000 m	Myriametric (VLF)
1000–10 000 m	Kilometric (LF)
100–1000 m	Hectometric (MF)
10–100 m	Decametric (HF)
1–10 m	Metric (VHF)
0.1–1 m	Decimetric (UHF)
1–10 cm	Centimetric (SHF)
0.1–1 cm	Millimetric (EHF)
0.01–0.1 cm	Decimillimetric (EHF)

The kilometric band is sometimes referred to as long wave (LW), the hecto-metric band as medium wave (MW) and the decametric band as short wave (SW). The metric and decimetric bands are known as ultra short wave (USW), whilst the centimetric and millimetric bands are generally referred to as microwaves.

Microwave band designations

The microwave bands were originally used by the military for radar systems, and the various bands of frequencies used were given single-letter code names. Although the letter sequence is similar, the corresponding frequency bands differ between Europe and the USA. Some of these band names are also used for satellite broadcasting but again the frequency ranges differ slightly from the military definitions.

European (NATO) microwave bands

In Europe the NATO forces adopted the following names for microwave bands:

1.0–2.0 GHz	L band
2.0–4.0 GHz	S band
4.0–8.0 GHz	C band
7.0–12 GHz	X band
12–18 GHz	J band
18–26 GHz	K band
26–40 GHz	Q band
40–60 GHz	V band
60–90 GHz	O band

USA microwave bands

In America a different definition is used for the microwave bands as follows:

0.4–1.5 GHz	L band
1.5–5.2 GHz	S band
3.7–6.2 GHz	C band
5.2–10.9 GHz	X band
10.9–36 GHz	K band
36–46 GHz	Q band
46–56 GHz	V band
56–100 GHz	W band

Satellite broadcast and communications bands generally follow these American microwave band definitions. The K band is subdivided into Ku and Ka bands.

CCITT regions
World-wide administration of radio is divided up into three major regions as follows:

Region 1   Europe, Africa and all of Russia
Region 2   Asia and Oceania
Region 3   North and South America

Frequency allocations and band usage vary slightly between the three regions.

General frequency allocations
From time to time a world-wide radio administration conference (WARC) is called to decide on the allocation and uses of the various frequency bands in the radio spectrum. The following list shows the current allocations to the various services.

VLF, LF, MF bands (frequency in kHz)
10.0–140.5	Fixed, maritime, navigation
140.5–283.5	Broadcast (AM)
255.0–526.5	Radionavigation, fixed
526.0–1606.5	Broadcast (AM regions 1 and 3)
526.0–1705.0	Broadcast (AM region 2)
1606.5–1800.0	Maritime and land mobile, fixed
1800–2000	Amateur
1850–2045	Fixed, mobile
2045–2173.5	Maritime mobile, fixed
2160–2170	Radiolocation
2173.5–2190.5	Mobile
2190.5–2194	Maritime
2194–2625	Fixed, mobile
2300–2498	Broadcast (AM tropical zone)
2498–2502	Frequency and time stations
2625–2650	Maritime mobile
2650–2850	Fixed, mobile
2850–3155	Aircraft

HF bands (frequency in kHz)
3155–3400	Fixed, mobile
3200–3400	Broadcast (AM tropical zone)
3400–3500	Aircraft
3500–3800	Amateur (shared)

3800–4000	Amateur (region 2 only)
3500–3900	Fixed, mobile
3800–3950	Aircraft
3950–4000	Fixed, broadcast
3950–4000	Broadcast (AM regions 1 and 3)
4000–4210	Maritime ships
4210–4438	Maritime shore
4438–4650	Fixed, mobile
4650–4700	Aircraft
4700–4750	Aircraft military
4750–5060	Broadcast (AM tropical zone)
4750–5480	Fixed, land mobile
4995–5005	Frequency and time stations
5450–5680	Aircraft civil
5680–5730	Aircraft military
5730–5950	Fixed, mobile
5900–5950	Broadcast (SSB)
5950–6200	Broadcast (AM)
6200–6314	Maritime ships
6314–6525	Maritime shore
6525–6685	Aircraft civil
6685–6765	Aircraft military
6765–700	Fixed, land mobile
7000–7100	Amateur
7100–7300	Amateur (region 2 only)
7100–7300	Broadcast (AM regions 1 and 3)
7300–7350	Broadcast (SSB)
7300–8100	Fixed, land mobile
8100–8417	Maritime ships
8417–8815	Maritime shore
8815–8965	Aircraft civil
8965–9040	Aircraft military
9040–9500	Fixed
9400–9500	Broadcast (SSB)
9500–9900	Broadcast (AM)
9900–9995	Fixed
9995–10 005	Frequency and time stations
10 005–10 100	Aircraft
10 100–11 175	Fixed
10 100–10 150	Amateur
11 175–11 275	Aircraft military
11 275–11 400	Aircraft civil
11 400–11 650	Fixed
11 600–11 650	Broadcast (SSB)

11 650–12 050	Broadcast (AM)
12 050–12 230	Fixed
12 100–12 230	Broadcast (SSB)
12 230–12 580	Maritime ships
12 580–13 200	Maritime shore
13 200–13 260	Aircraft military
13 260–13 360	Aircraft civil
13 360–13 410	Fixed
13 410–13 870	Fixed, land mobile
13 570–13 600	Broadcast (SSB)
13 600–13 800	Broadcast (AM)
13 800–13 870	Broadcast (SSB)
13 800–14 000	Fixed, land mobile
14 000–14 350	Amateur
14 350–14 990	Fixed, land mobile
14 990–15 010	Frequency and time stations
15 010–15 100	Aircraft military
15 100–15 600	Broadcast (AM)
15 600–15 800	Broadcast (SSB)
15 600–16 360	Fixed
16 360–16 807	Maritime ships
16 807–17 410	Maritime shore
17 410–17 550	Fixed
17 480–17 550	Broadcast (SSB)
17 550–17 900	Broadcast (AM)
17 900–17 970	Aircraft civil
17 970–18 030	Aircraft military
18 030–18 068	Fixed
18 068–18 168	Amateur
18 168–18 780	Fixed, land mobile
18 780–18 900	Maritime ships
18 900–19 020	Broadcast (SSB)
18 900–19 680	Fixed
19 680–19 800	Maritime shore
19 800–19 900	Fixed
19 990–20 010	Frequency and time stations
20 010–21 000	Fixed, land mobile
21 000–21 450	Amateur
21 450–21 850	Broadcast (AM)
21 850–21 870	Fixed
21 870–22 000	Aircraft civil
22 000–22 376	Maritime ships
22 376–22 855	Maritime shore
22 855–23 000	Fixed

23 000–23 200	Fixed, land mobile
23 200–23 350	Aircraft military
23 350–24 890	Fixed, land mobile
24 890–24 990	Amateur
24 990–25 010	Frequency and time stations
25 010–25 070	Fixed, land mobile
25 070–25 210	Maritime ships
21 210–25 550	Fixed, mobile
25 550–25 670	Radio astronomy
25 670–26 100	Broadcast (AM)
26 100–26 175	Maritime shore
26 175–28 000	Fixed, mobile
26 965–27 405	CB radio (FM CEPT channels)
27 601–28 000	CB radio (FM UK channels)
26 965–28 000	CB radio (AM/SSB North America)
28 000–29 700	Amateur
29 700–30 000	Fixed, mobile

VHF and UHF bands (frequencies in MHz)

30.0–50.0	Fixed, mobile
35.0–35.25	Model control
47.0–68.0	Broadcast (TV)
47.0–50.0	Mobile and paging systems
50.0–52.0	Amateur (UK)
50.0–54.0	Amateur (regions 2 and 3)
52.0–68.0	Mobile
68.0–70.8	Mobile (mainly military)
70.0–70.5	Amateur (UK)
70.5–74.8	Mobile
74.8–75.2	Aero navigation (ILS)
75.2–88.0	Fixed, mobile
88.0–108.0	Broadcast (FM)
108.0–118.0	Aero navigation (VOR)
118.0–137.0	Aircraft civil
136.0–138.0	Satellite downlink
138.0–144.0	Land mobile
144.0–146.0	Amateur (region 1)
144.0–148.0	Amateur (regions 2 and 3)
146.0–149.9	Fixed and mobile
149.9–150.1	Satellite navigation
150.1–152.0	Radio astronomy
152.0–174.0	Fixed, land mobile
156.0–162.0	Maritime ships and shore
162.0–225.0	Fixed, mobile (UK)

174.0–225.0	Broadcast (TV) Band 3 (not UK)
220.0–225.0	Amateur (USA)
225.0–328.6	Aircraft military
328.6–335.4	Aero navigation (ILS)
335.4–399.9	Aircraft military
399.9–400.1	Satellite navigation
400.1–410.0	Space research, meteorology
406.0–410.0	Radio astronomy
410.0–430.0	Fixed, mobile
430.0–440.0	Amateur, radiolocation
440.0–450.0	Mobile
450.0–470.0	Fixed, mobile
470.0–855.0	Broadcast (TV)
855.0–960.0	Fixed, mobile
902.0–928.0	Amateur (USA)
934.0–935.0	CB (UK)
960.0–1215	Aero navigation (DME)
1215–1240	Satellite navigation, radar
1240–1325	Amateur
1300–1350	Aero navigation
1350–1400	Fixed, mobile
1400–1429	Space (uplink)
1429–1525	Fixed, mobile
1525–1600	Space (downlink) L band
1600–1670	Space (uplink)
1610–1627	Space uplink mobile radio
1670–1710	Space (downlink)
1710–2290	Fixed, mobile
2290–2300	Space downlink
2300–2450	Amateur, fixed
2310–2450	Amateur (UK)
2300–2700	Fixed, mobile
2483–2520	Space downlink mobile radio
2500–2700	Space downlink S band
2700–3300	Radar
3300–3400	Radiolocation, amateur
3400–3600	Fixed
3400–4200	Space downlink C band
4200–4400	Aero navigation
4400–5000	Fixed, mobile
4500–4800	Space downlink C band
4800–5000	Fixed, mobile
5000–5850	Radio navigation, radar
5650–5850	Amateur

5855–6055	Space uplink L band
5725–6425	Space uplink C band
5850–7250	Fixed
6425–7075	Space uplink C band
7250–7750	Space downlink X band
7900–8400	Space uplink X band
7250–8500	Fixed, mobile
8500–1050	Radar, navigation
10 000–10 500	Amateur
10 700–11 700	Space downlink Ku FSS band
11 700–12 500	Space downlink Ku BSS band
12 500–12 750	Space downlink Ku FSS band
12 750–13 250	Space uplink Ku FSS band
14 000–14 250	Space uplink Ku BSS band
14 000–14 500	Space uplink Ku FSS band
17 300–18 100	Space uplink Ku DBS band
18 300–22 200	Space downlink Ka band
27 000–31 000	Space uplink Ka band
24 000–24 250	Amateur
47 000–47 200	Amateur
75 500–76 000	Amateur
142 000–144 000	Amateur
248 000–250 000	Amateur

The SSB segments of the HF broadcast bands will not become fully available until 2007 but it is likely that some stations will start operation before that date.

Emission designations

There are a number of different ways in which information can be conveyed on a radio signal. The signal may be varied in amplitude, frequency or phase, and the actual modulation used may be either analogue or digital or a combination of both. In the past the signal would be described in terms such as amplitude modulation (AM), frequency modulation (FM) and single sideband (SSB). In order to define exactly every possible type of transmission, the CCIR (Consultative Committee International Radio) has developed a comprehensive set of emission designation codes. The standard emission definitions adopted by the CCIR consist of up to five symbols, although for most applications only the first three are normally used. The first symbol indicates the type of modulation being used. The second symbol is a number which indicates the type of modulating signal being used, and the third symbol is a letter which indicates the type of information being conveyed by the modulating signal. These are specified as follows:

First symbol	Type of modulation of carrier
N	Unmodulated carrier
A	Double sideband (AM)
H	Single sideband full carrier
J	Single sideband suppressed carrier
R	Single sideband reduced carrier
B	Independent sideband modulation
C	Vestigial sideband AM
F	FM
G	Phase modulation
P	Unmodulated pulses
K	Amplitude-modulated pulses
L	Width- or duration-modulated pulses
M	Position- or phase-modulated pulses
X	Cases not otherwise covered

Second symbol	Type of modulating signal
0	No modulating signal
1	Single digital channel
2	Single digital channel using subcarrier
3	Single analogue channel
7	Multichannel digital signal
8	Multichannel analogue signal
9	Combination of 7 and 8

Third symbol	Type of information
N	No information
A	Telegraphy (Morse)
B	Automatic telegraphy (RTTY etc.)
C	Facsimile
D	Data transmission
E	Telephony
F	Television (includes SSTV)
W	Television with sound
X	Other signals

The fourth symbol is used to define the type of coding used for digital signals (letters A to F) or the type of sound or vision transmission (i.e. monophonic or stereophonic sound, monochrome or colour video) (letters G to X).

The fifth symbol is used to indicate if the channel is multiplexed and the type of multiplexing system used.

For most applications only the first three symbols of the emission designation are used. Commonly used modes are:

A1A   Morse using on–off keying (CW)
A2A   Morse using a keyed tone (MCW)
A1B   RTTY using on–off keying
A2B   RTTY using on–off keyed tone

A3C   FAX using AM
A3E   AM telephony
A3F   Television or SSTV using AM
C3F   Television AM vestigial sideband
J3E   Single sideband suppressed carrier
F1A   Telegraphy by frequency shift keying (FSK)
F1B   Frequency shift keyed radioteletype
F2A   FM telegraphy by keyed audio tone
F2B   FM RTTY with FSK audio subcarrier (AFSK)
F3C   FAX using frequency modulation
F3E   FM telephony
F3F   Television or SSTV using FM
G1B   RTTY using phase shift keying
G2B   RTTY with phase shift subcarrier
G3E   Phase-modulated telephony
PON   Unmodulated pulse transmission
R3E   Reduced carrier mode telephony

2

Amateur radio licences

In most countries of the world licences for the operation of amateur radio stations are available to suitably qualified citizens and visitors. The requirements vary between countries, but usually involve a written technical examination and, for HF operation, some form of Morse test. Some countries have various grades of amateur licence which may include a novice licence for beginners in the hobby. Many countries will grant a reciprocal licence to visitors who hold an amateur licence in another country.

UK class A and B licences
The class A licence requires the operator to have passed the Radio Amateur's Examination (RAE) and to have passed a Morse test demonstrating the ability to send and receive Morse code at 12 words per minute. This licence permits the user to operate on all bands allocated for amateur radio use with all modes of transmission, including telegraphy, telephony, radioteletype (RTTY), facsimile (FAX), TV and data transmissions.

The class B licence also requires a pass in the RAE but no Morse test is needed. This licence permits operation of a transmitter on all amateur radio bands above 30 MHz using all modes of transmission permitted by the class A licence. If the holder of a class B licence goes on to take the Morse test, then a class A licence may be issued, allowing the station to operate on the HF amateur bands.

A class B licence holder is allowed to use the station of a class A licensee to operate on the HF bands provided that this is done under the supervision of the class A licensee and the callsign used is that of the class A station. The class B licence holder must then sign the station log to show that he or she was operating the station.

For the full class A or class B licence the applicant must be at least 14 years of age.

UK novice licences

In 1991 two new novice licences were introduced to encourage younger entrants into the amateur radio hobby. The novice licence is also attractive to older applicants who would like to try the hobby but are put off by the examination or Morse requirements for the full licence. There are two types of novice licences, which are known as Novice A and Novice B.

It seems likely that most novice licence holders will eventually go on to obtain a full class A or B licence in order to gain access to additional bands, modes of operation and the higher power levels permitted by the full licence.

To obtain a novice licence the applicant must first take one of the novice training courses organized by the Radio Society of Great Britain (RSGB). These are usually run by local amateur radio clubs by an instructor who has been approved by the RSGB. The course normally lasts for about 30 hours and is spread over a period of 12 weeks. At the end of the course the work done by the candidates is assessed and they are issued a completion slip by the RSGB. The course is a practical introduction to amateur radio, covering 32 topics and including some construction projects and the completion of a set of worksheets.

The topics covered in a novice course are:

Colour codes
Soldering exercises 1 and 2
Ohm's law
Codes and abbreviations
Audio amplifier construction project
Setting up a radio contact
Making logs and designing QSL cards
Introduction to aerials
Using a multimeter
Measuring resistance and current
Principles of radio propagation
QSL bureaux
AC signals and frequency
Tuned circuits
Fitting mains plugs
The radio spectrum
Fitting BNC and PL259 plugs to coaxial cables
Block diagrams of receivers and transmitters
Harmonics
Learning the Morse code
Preparing for a Morse contact
Electromagnetic compatibility (interference)
Power supplies

Practical work includes the construction of a test set for demonstrating Ohm's law and for tests on diodes and transistors.

Courses are usually limited to three or four students so that the instructor can give individual attention. Once the course is complete the applicant must also take a novice examination organized by the City and Guilds Institute. This is a multiple choice paper similar to the RAE, with 45 questions based on topics covered in the novice training course. This examination takes place at a number of centres around the country and the course instructor will have details of the local venue. Results are usually available about a month after the examination date.

For a Novice B licence the completion slip for the training course and a pass slip for the City and Guilds examination are sufficient qualification and these can be sent off with the fee of £15 to obtain the licence.

For a Novice A licence the applicant must also pass a Morse code test at five words per minute, which is not difficult to achieve. The Morse tests are organized by the Radio Society of Great Britain (RSGB) at a number of centres around the country. Tests are usually held every two months and consist of a three-minute session of transmitting Morse followed by a six-minute session of receiving. The test pieces consist of a mixture of plain text, numbers and procedural codes which simulate the sort of signals which might be exchanged during a typical amateur radio contact.

The Novice B licence permits the user to operate on some of the VHF and UHF bands available to the full class B station but with a power limit of 3 W output. This permits the use of CW, data (packet radio) and phone operation on all of the allocated bands and all modes on two of the higher UHF bands.

For the class A novice, parts of the HF bands are available which would permit world-wide contacts under favourable conditions. On most of the HF bands only Morse is permitted. On the 1.8-MHz and 28-MHz bands Novice A stations are allowed to use RTTY, packet radio and telephony modes. Class A novices can also use all of the VHF and UHF bands and modes allocated to the Novice B licence.

The schedule of frequencies and transmission modes available to novice operators is given in the next chapter.

Radio amateur's examination

This is a technical examination organized by the City and Guilds Institute of London, which examines the applicant for a basic knowledge of the principles involved in radio transmission and reception and the rules and conditions of the amateur radio licence. Particular emphasis is placed on the knowledge of operation of equipment in a proper manner to avoid causing interference to other services.

The examination itself takes the form of a set of two multiple choice papers in which three or four possible answers are given for each question.

The student is then required to select the most appropriate answer for each question. This form of paper has the advantage that it can readily be marked by using computer techniques, thus reducing the time needed to evaluate the examination results. One part of the examination is concerned with knowledge of the rules and conditions for amateur radio whilst the other is a technical paper covering basic aspects of radio equipment and its operation. The examination takes place twice a year at a large number of centres throughout the UK.

Morse tests

A class A licence applicant is required to pass a test demonstrating the ability to send and receive Morse at 12 words per minute. Tests are organized by the RSGB on behalf of the Department of Trade and Industry. The RSGB Morse tests are arranged at many locations in the UK throughout the year. Facilities for taking the Morse test are often provided at amateur radio rallies and conventions. Bookings for Morse tests are made in advance of the test date and applicants should contact the RSGB for a list of future test venues and dates in order to book a Morse test.

For the test, about 36 words of plain language text, often taken from a book or magazine, are sent using a standard Morse key connected to a buzzer or tone generator. Up to four corrections may be made whilst sending, and all errors must be corrected. A similar plain language text must then be received and written on paper with less than four errors. The time allowed for each of these tests is three minutes. A further test involving sending and receiving 10 groups of five figures is then made. Here the time for each test is 1½ minutes and only two errors or corrections are permitted. If the applicant satisfies the examiner that he or she is able to send and receive Morse correctly at the required speed, a pass certificate is awarded.

US amateur licences

In the USA there are five classes of amateur licence known as Novice, Technician, General, Advanced and Extra. The type of licence held determines the bands available, permitted modes of operation and maximum power that can be used by the station.

The lowest class is the Novice, which requires the holder to pass a relatively simple technical examination and be able to send and receive Morse code at five words per minute. Novices may use CW-only operation on limited sections of the 3.5-, 7-, 21- and 28-MHz bands with a maximum of 200 W input power.

The Technician class licencee has to pass a more difficult technical examination. Once again the ability to send and receive Morse dode at five words per minute is required. This licence allows use of all modes on the VHF and UHF bands as well as CW in the Novice sections of the HF bands.

The standard licence is the General licence, which has the same technical examination as that for the Technician licence. For the Morse test a speed of 13 words per minute, sending and receiving, is required. The General licence permits the use of phone and CW on segments of all of the HF amateur bands with 1 kW input power. In addition, all modes may be used on VHF and UHF, again with 1 kW input power.

By passing a further technical examination the amateur can obtain an Advanced licence which permits the use of higher power and more modes on the HF bands. The higher licence class is Extra, giving all modes and bands.

Third-party traffic

Some countries, such as the USA, allow their amateur radio operators to handle messages from members of the public who are not licensed amateurs. In fact, many US amateurs handle radiogram messages on a non-commercial basis as a regular part of their amateur radio activity. Another regular activity is that of 'phone patching', where a local member of the public can be linked to someone at a distance via amateur radio. Here the caller telephones the amateur at local call rates and the amateur tries to establish contact with another amateur in the town where the caller wishes to make a call. The second amateur then makes the local phone call and the phone conversation is then patched through the amateur radio link. This mode is often used by US servicemen in Europe for making calls home to their families. This form of operation is known as handling third-party traffic.

In the UK the licence does not permit the handling of messages for third parties by normal amateur stations. The exception is special event stations with GB callsigns where it is permissible for a non-licensed person to send a simple greetings message to the station which is being contacted. This greetings message concession is used at events such as the Scouts jamboree on the air. In an emergency, amateur stations may handle messages for the emergency services.

An exception to this rule applies when the station is operated as a node in the packet radio network where forwarding of packets to the next node or station in the link is allowed.

Reciprocal licences

Reciprocal licensing agreements with many countries allow licensed amateurs from those countries to operate in the UK. These stations may be granted either class A or class B status depending upon the conditions associated with the licence held in their own country. The callsign used by such stations is normally a G prefix of the appropriate type followed by their own home callsign (e.g. G0/W2ABC).

Under these reciprocal arrangements a UK amateur may obtain a licence to operate in foreign countries such as the USA and most European countries. The application for a reciprocal licence usually needs to be made well in advance of a visit, and is normally made to the licensing authority of the country concerned. More details of reciprocal licensing arrangements may be obtained from the RSGB.

The CEPT licence

In Europe a number of countries have a reciprocal arrangement known as the CEPT licence. The countries which are currently operating this agreement are:

Austria	Luxembourg
Belgium	Malta
Cyprus	Monaco
Denmark	Netherlands
Finland	Norway
France	Portugal
Germany	San Marino
UK	Spain
Greece	Sweden
Iceland	Switzerland
Ireland	Turkey
Italy	Vatican City
Liechtenstein	Yugoslavia

The holder of a permanent licence in any of these countries may operate an amateur radio station in any of the other countries. This does not apply to novice licences. The licence holder must be a temporary visitor to the other country in which the station is to be operated. The UK licence or validation document should also be available for inspection by the local authorities if requested.

The station must be either mobile or portable, which may include operation of the equipment from a mains supply at a temporary location such as a hotel. The visiting amateur may also operate the fixed station of a radio amateur who is resident in the country being visited.

Note that the frequency bands which may be used are limited to those permitted in the country being visited. Thus although the 50- and 70-MHz bands are available to UK amateurs, these bands are not allocated for amateur use in many of the CEPT countries and therefore may not be used by the visiting UK amateur. Class B licensees are only permitted to use VHF and UHF, which will generally mean 144 MHz and above.

The power output and emission modes should comply with the home licence unless the licence conditions in the country call for lower power limits or do not permit certain modes of operation.

The callsign used when operating in another CEPT country consists of the prefix of the host country followed by the home callsign. Thus a UK amateur operating in Malta might use the callsign 9H/G3ABC if the home callsign is G3ABC.

The Morse code

This code was originally devised for use on commercial telegraph circuits and is still widely used for manual telegraphy. Special versions of the code may be used in countries, such as Japan, where a different alphabet is used. The international version of the Morse code is:

A	._	N	_.	1	._ _ _ _
B	_...	O	_ _ _	2	.._ _ _
C	_._.	P	._ _.	3	..._ _
D	_..	Q	_ _._	4	...._
E	.	R	._.	5	.....
F	.._.	S	...	6	_....
G	_ _.	T	_	7	_ _...
H	....	U	.._	8	_ _ _..
I	..	V	..._	9	_ _ _ _.
J	._ _ _	W	._ _	0	_ _ _ _ _
K	_._	X	_.._		
L	._..	Y	_._ _		
M	_ _	Z	_ _..		

Special symbols

Query	(?)	.._ _..
Stop	(.)	._._._
Hyphen	(-)	_...._
Comma	(,)	_ _.._ _
Slash	(/)	_.._.
Error		....

Procedure codes (letter pairs sent as a single code)

Wait	(AS)	._...
End of contact	(SK)	..._._
Go ahead	(K)	_._
Over to specific station	(KN)	_._ _.
Over when calling station	(AR)	._._.
Roger (all received OK)	(R)	._.
Closing down	(CL)	_._.._..

Morse code timing

For proper readability of Morse code it is important that the relative timing of the dots, dashes and pauses should be consistent throughout a message. All timing is related to the time period for a dot.

Dot time	1 dot
Dash time	3 dots
Space between elements	1 dot
Space between characters	3 dots
Space between words	7 dots

Morse code transmission speed is normally measured in words per minute. Speed calculations are based on an average word length of five characters and an average character length of 50-dot periods. Approximate dot times for typical Morse speeds are:

Words per min	Dot time (ms)
6	200
12	100
18	67
24	50

Most amateur radio licence rules require a morse speed of around 12 words per minute for an HF band licence. Some countries also have novice licences for which a speed of about 5 words/min is usually required. Novice stations usually have restricted access to the HF bands and lower power limits.

Morse practice

Many amateur radio clubs run classes to teach newcomers how to send and receive Morse code in preparation for the Morse test. Often a local amateur will arrange to send practice Morse transmissions to club members so that they can develop their ability to read Morse code from an off-air signal. In Britain the class B stations are permitted to send and receive Morse signals on the VHF bands as part of their training to upgrade to a full class A licence. Practice nets are sometimes arranged by clubs to allow class B stations in the area to practise sending and receiving Morse code.

There are a number of regular broadcast transmissions of slow-speed Morse for listeners and amateurs wishing to improve their ability to receive Morse code. These transmissions cover a range of speeds and normally include both plain language passages and mixed figure/letter groups.

Morse practice transmissions in the UK

Transmissions on HF and VHF are organized by local amateur radio clubs. Some regular HF band transmissions are:

G3GNS	RAF Locking, Avon	1910 kHz, 3550 kHz
	Mo We Th Fr 1830	
	Tu Sa Su 1200	
G4RS	Catterick, North Yorkshire	3565 kHz
	Tu Th 1900	
G4PYR	Solihull, West Midlands	1888 kHz
	Su Mo 1900	
G4OBK	Chorley, Lancashire	3565 kHz
	Fr 1930	
GM4HYF	Glasgow	28 350 kHz
	We 2130 Th 2200	

Most VHF transmissions use 144.250 MHz (A1A) or 145.250 MHz (F2A) at various times between 1830 and 2200.

A full list of Morse practice transmissions for listeners in the UK is available from the RSGB.

Morse practice transmissions from Veron, Holland
Station PAOAA

3602 kHz, 14103 kHz, 144.800 MHz, 433.45 MHz
Fridays only
1900–1930   For beginners
1930–2000   For advanced listeners

Morse practice transmissions from the USA
These transmissions are broadcast from the ARRL Headquarters Station W1AW in Newington Connecticut on frequencies of 1818, 3580, 7080, 14 070, 21 080, 28 080 and 50 080 kHz.

5 to 15 words per minute
Mo We Fr   1400, 2400 GMT
Tu Th Sa Su   2100 GMT
We Fr Su Mo   0300 GMT

10 to 35 words per minute
Tu Th   0300, 1400, 2400 GMT
Sa Su   2400 GMT
Mo   2100 GMT

Times are one hour earlier during summer months.

3

Amateur radio bands

Many small sections of the radio spectrum are allocated for use by radio amateurs. The frequencies available vary slightly between the three CCITT regions. On some amateur bands the amateur stations are primary users, whilst on many bands they operate as secondary users. Amateur users are, however, not protected from interference by other users even on bands where they are the primary users.

Amateur radio bands in the UK

MF band (class A stations only)

Frequency (kHz)	Peak envelope power (PEP) output (W)	Status
1810–1830	32	Primary
1830–1850	400	Primary
1850–2000	32	Secondary

HF bands (class A stations only)

Frequency (kHz)	PEP output (W)	Status
3500–3800	400	Primary
7000–7100	400	Primary
10 100–10 150	400	Secondary
14 000–14 350	400	Primary
18 068–18 168	400	Primary
21 000–21 450	400	Primary
24 890–24 990	400	Primary
28 000–29 700	400	Primary

The emission modes permitted on these MF and HF bands are:

Morse   A1A, A2A, F1A, F2A, G1A, G2A
RTTY   A1B, A2B, F1B, F2B, G1B, G2B
Packet   F1D, F2D, G1D, G2D, J2D
Telephony   A3E, J3E, F3E, G3E
FAX   A3C, F3C, G3C, J2C
SSTV   A3F, C3F, J2F

On the 10-MHz band, although all modes are permitted, it is recommended that only the narrow-band modes (Morse, RTTY, packet) should be used at the present time because of the limited space available.

VHF bands (class A and class B stations)

Frequency (MHz)	SSB PEP output (W)	Status
50.0–51.0	100 W erp	Primary
51.0–52.0	100 W erp	Secondary
70.0–70.5	160	Secondary
144–146	400	Primary

Note: erp is the effectvie radiated power which includes the antenna power gain

The emission modes permitted on these bands are:

Morse   A1A, A2A, F1A, F2A, G1A, G2A
RTTY   A1B, A2B, F1B, F2B, G1B, G2B
Packet   F1D, F2D, G1D, G2D, J2D
Data   F1D, F2D, G1D, G2D, J2D
Telephony   A3E, J3E, F3E, G3E
FAX   A3C, F3C, G3C
SSTV   A3F, C3F

UHF bands (class A and class B stations)

Frequency band (MHz)	SSB PEP output (W)	Status
430–432	40 W erp	Secondary
432–440	400	Secondary
1240–1325	400	Secondary
2310–2450	400	Secondary

The emission modes permitted on these bands are:

Morse   A1A, A2A, F1A, F2A, G1A, G2A

RTTY	A1B, A2B, F1B, F2B, G1B, G2B
Packet	F1D, F2D, G1D, G2D, J2D
Data	F1D, F2D, G1D, G2D, J2D
Telephony	A3E, J3E, F3E, G3E
FAX	A3C, F3C, G3C
SSTV	A3F, C3F
FSTV	A3F, C3F, F3F, G3F

SHF and EHF bands (class A and class B stations)

Frequency	SSB PEP output (W)	Status
3400–3475 MHz	400	Secondary
5650–5680 MHz	400	Secondary
5755–5765 MHz	400	Secondary
5820–5850 MHz	400	Secondary
10.00–10.50 GHz	400	Secondary
24.00–24.05 GHz	400	Primary
24.05–24.25 GHz	400	Secondary
47.00–47.20 GHz	400	Primary
75.50–76.00 GHz	400	Primary
142.0–144.0 GHz	400	Primary
248.0–250.0 GHz	400	Primary

Use of the band 24.05–24.15 GHz requires written consent from the Secretary of State at the Department of Trade and Industry.

Users of the 2.4-, 2.45-, 5.7-, 5.8- and 24-GHz bands must accept possible interference from industrial, scientific and medical equipment users.

The band 431–432 MHz may not be used within 100 km of Charing Cross in London.

On the 50-MHz band antenna height is limited to 20 m above ground level.

The emission modes permitted on these bands are:

Morse	A1A, A2A, F1A, F2A, G1A, G2A
RTTY	A1B, A2B, F1B, F2B, G1B, G2B
Packet	F1D, F2D, G1D, G2D, J2D
Data	F1D, F2D, G1D, G2D, J2D
Telephony	A3E, J3E, F3E, G3E
FAX	A3C, F3C, G3C
SSTV	A3F, C3F, J2F
FSTV	A3F, C3F, F3F, G3F

Pulse modes are permitted on the 2.4-, 5.7-, 5.8- and 10.0-GHz bands and on 47, 75, 142 and 250 GHz.

MF band for UK Novice A licence

Frequency (kHz)	PEP output (W)	Mode
1950–2080	3	Morse, RTTY, data, phone

HF bands for UK Novice A licence

Frequency band (kHz)	PEP output (W)	Mode
3565–3585	3	Morse
10 130–10 140	3	Morse
21 100–21 149	3	Morse
28 100–28 190	3	Morse, RTTY, data
28 225–28 300	3	Morse, RTTY, data
28 300–28 500	3	Morse, phone

The data mode includes packet radio transmissions.

VHF band for UK Novice A and B licences

Frequency (MHz)	PEP output (W)	Mode
50.00–52.00	3	Morse, data, phone

UHF bands for UK Novice A and B stations

Frequency (MHz)	PEP output (W)	Mode
432.0–435.0	3	Morse, data, phone
435.0–440.0	3	Morse, data, phone, SSTV, FSTV
1240–1325	3	Morse, data, phone, FAX, RTTY, SSTV, FSTV

SHF band for UK Novice A and B stations

Frequency (MHz)	PEP output (W)	Mode
10 000–10 500	3	Morse, data, phone, FAX, RTTY, SSTV, FSTV

Data includes packet radio.

On the 50-MHz band antenna height is limited to 20 m above ground level.

Amateur MF and HF band plans

By general agreement among radio amateurs, the amateur bands have been divided into segments which are preferred areas for various modes of transmission. In some countries, such as the USA, the types of transmission allowed in various segments of the bands are laid down by the licence. In general the lower frequency part of a band is used for Morse, whilst the higher frequency part is used for telephony. Special modes, such as RTTY, packet and SSTV, tend to have small segments at the upper end of the Morse area or within part of the telephony (phone) section.

Note that some of the frequencies shown in these band plans are available only in North America.

160-metre band

1810–1838 kHz	Morse
1838–1842 kHz	Morse, RTTY
1842–2000 kHz	Morse, phone

80-metre band

3500–3600 kHz	Morse
3580–3620 kHz	RTTY
3620–3775 kHz	Morse, phone
3730–3740 kHz	SSTV
3775–3800 kHz	Dx phone
3800–4000 kHz	Phone (USA)

40-metre band

7000–7035 kHz	Morse
7035–7045 kHz	RTTY, SSTV
7045–7100 kHz	Phone
7080–7100 kHz	RTTY (USA)
7100–7150 kHz	Morse (USA)
7150–7300 kHz	Phone (USA)

30-metre band

10 100–10 140 kHz	Morse
10 140–10 150 kHz	Morse, RTTY, packet

20-metre band

14 000–14 060 kHz	Morse
14 060–14 110 kHz	Morse, RTTY, packet
14 110–14 350 kHz	Morse, phone
14 225–14 235 kHz	SSTV
14 240–14 250 kHz	FAX

17-metre band

18 068–18 100 kHz	Morse
18 100–18 110 kHz	Morse, RTTY, packet
18 110–18 165 kHz	Morse, phone

15-metre band

21 000–21 080 kHz	Morse
21 080–21 120 kHz	Morse, RTTY, packet
21 120–21 150 kHz	Morse
21 150–21 335 kHz	Phone
21 335–21 345 kHz	SSTV
21 345–21 450 kHz	Phone

12-metre band

24 890–24 920 kHz	Morse
24 920–24 930 kHz	RTTY, packet
24 930–24 990 kHz	Phone

10-metre band

28 000–28 050 kHz	Morse
28 050–28 150 kHz	RTTY, packet
28 150–28 190 kHz	Morse
28 190–28 300 kHz	Beacons
28 675–28 685 kHz	SSTV
28 685–29 300 kHz	Phone
29 300–29 550 kHz	Satellite band
29 510–29 590 kHz	FM repeater inputs
29 600 kHz	FM calling frequency
29 610–29 670 kHz	FM repeater outputs

Amateur VHF band plans in the UK (frequencies in MHz)

Six-metre band

50.00–50.00	Morse only
50.10–50.50	Morse/SSB
50.50–52.00	All modes
50.60–50.80	Packet

Four-metre band

70.025–70.075	Beacons only
70.000–70.150	Morse only
70.150–70.260	Morse/SSB
70.260–70.400	All modes

| 70.400–70.500 | FM simplex |
| 70.300–70.500 | Packet |

Two-metre band

144.00–144.15	CW only
144.15–144.50	CW/SSB only
144.50	SSTV
144.60	RTTY (FSK)
144.60–144.70	Packet
144.80–144.84	All modes
144.84–145.00	Beacons
145.00–145.20	FM repeater inputs
145.20–145.60	FM simplex channels
145.30	RTTY (AFSK)
145.50	FM calling channel
145.60–145.80	FM repeater outputs
145.80–146.00	Satellite band

Amateur UHF band plans in the UK (frequencies in MHz)

70-centimetre band

430.00–432.00	Morse only
430.60–431.00	Packet
432.00–432.15	Morse
432.15–432.50	Morse/SSB
432.50	SSTV
432.50–432.80	All modes
432.10–432.40	Packet
432.60	RTTY (FSK)
432.80–433.00	Beacons
433.00–433.40	FM repeater outputs
433.30	RTTY (AFSK)
433.40–434.60	FM simplex
434.60–435.00	FM repeater inputs
435.00–438.00	Satellite band
434.00–440.00	Amateur TV

23-centimetre band

1240.000–1256.000	Amateur TV
1240.150–1240.750	Packet
1256.000–1260.000	All modes
1260.000–1270.000	Amateur satellite
1270.000–1286.000	Amateur TV

1286.000–1291.000	All modes
1291.000–1291.475	Repeater inputs
1296.000–1296.025	Moonbounce
1296.025–1296.500	Narrow-band Dx operation
1296.500–1296.600	Linear transponder input
1296.600–1296.700	Linear transponder output
1296.800–1296.990	Beacons only
1297.000–1297.475	Repeater outputs
1297.500–1298.000	Simplex channels
1298.000–1300.000	All modes
1299.000–1300.000	Packet
1300.000–1325.000	TV repeater outputs

4

Mobile and portable operation

Although most amateur stations operate from their permanent home locations, the licence also permits mobile operation from a car or portable operation from a temporary site or whilst walking. The licence also permits operation from a boat on inland waterways or lakes. An additional maritime mobile licence may be obtained which permits operation from a boat or ship at sea. In some countries the station may be operated from an aircraft or balloon, but this concession is not available in the UK.

Mobile and portable callsigns

The UK class A or B licence permits both mobile and portable operation anywhere in the UK. When the station is operated from a vehicle, the suffix /M is added after the callsign (e.g. G3ABC/M). For a station operated as a portable transmitter, the suffix /P is added to the callsign. The /P is also used when the station is set up at a temporary address different from that listed on the licence.

The callsign prefix changes according to the part of the UK that the station is operating from. A second letter following the G, M or 2 indicates the location. The secondary location letters are:

D   Isle of Man
I   Northern Ireland
J   Jersey
M   Scotland
U   Guernsey
W   Wales

Novice stations use the call 2E when based in England, whilst class A and B stations use the G or M prefix with no secondary location letter. The country prefix is also used by stations which are permanently located in that country. Thus a G station which moved to Scotland would then use a GM prefix.

Club stations operating portable or mobile transmitters use a different set of secondary location identifier letters as follows:

C   Wales
H   Jersey
N   Northern Ireland
P   Guernsey
S   Scotland
T   Isle of Man
X   England

For example, GM3AAA/M would be a mobile station in Scotland, 2W1ZZZ/P would be a novice operating a portable station in Wales and GX3CCC/P would be a club portable station operating in England. Note that the club station prefixes are also used when the club station operates from its permanent location.

The ordinary class A or B licence may be used with a /M prefix for operation from a boat on inland waterways or lakes but operation is not permitted at sea or in a harbour or estuary. For true maritime mobile operation at sea or in a harbour a separate licence is required. The callsign used is the same as that allocated for the class A or B licence but during maritime mobile operation the suffix /MM is added. The suffix /MA may be used when the boat or ship is anchored. For this type of operation from a commercial ship the permission of the master and owners of the ship must be obtained. In the case of a privately owned yacht or boat this is normally not a problem, since the licensee is usually the owner.

Some countries, such as the USA, permit operation from aircraft as an aeronautical mobile with the suffix /AM. This mode of operation is not permitted by the amateur radio licence in the UK.

VHF FM simplex channels in the UK

On the VHF and UHF bands a section of each band has been channelized for simplex operation using narrow-band FM with 5-kHz deviation. These channels may be used by fixed, mobile or portable stations which transmit and receive on the same frequency. On each band one frequency is designated as a calling channel. This channel is used to establish initial contact and then the frequency is changed to any one of the other channels that is not currently occupied.

On the 144-MHz band there are 16 channels, spaced 25 kHz apart, which are numbered S8 to S23.

Channel No.	Frequency (MHz)
S8	145.200
S9	145.225
S10	145.250
S11	145.275
S12	145.300 RTTY (AFSK)
S13	145.325
S14	145.350
S15	145.375
S16	145.400
S17	145.425
S18	145.450
S19	145.475
S20	145.500 FM calling channel
S21	145.525 RSGB news (Sundays)
S22	145.550 Talk-in for rallies
S23	145.575

UHF FM simplex channels in the UK

On the 432-MHz band there are 14 simplex channels which are spaced 25 kHz apart and numbered SU16 to SU20. An additional channel (SU12) is used for RTTY.

Channel No.	Frequency (MHz)
SU16	433.400
SU17	433.425
SU18	433.450
SU19	433.475
SU20	433.500 FM calling channel
SU21	433.525
SU22	433.550 Talk-in for rallies
SU23	433.575
SU24	433.600 RTTY (AFSK)
SU25	433.700 RAYNET
SU26	433.725 RAYNET
SU27	433.750 RAYNET
SU28	433.775 RAYNET
SU12	433.300 RTTY (AFSK)

On the 1.3-GHz band 21 simplex channels at 25-kHz spacing have been allocated. These are numbered from SM20 to SM40.

Channel No.	Frequency (MHz)
SM20	1297.500 Calling channel
SM21	1297.525
SM22	1297.550
SM23	1297.575
SM24	1297.600
SM25	1297.625

Channel No.	Frequency (MHz)
SM26	1297.650
SM27	1297.675
SM28	1297.700
SM29	1297.725
SM30	1297.750
SM31	1297.775
SM32	1297.800
SM33	1297.825
SM34	1297.850
SM35	1297.875
SM36	1297.900
SM37	1297.925
SM38	1297.950
SM39	1297.975
SM40	1298.000

FM repeaters

For mobile radio operation amateur clubs and groups have set up a number of repeater stations which permit the mobile station to achieve much greater coverage than by using simplex transmission. Repeaters are usually located on hills or tall buildings to give a wide coverage area.

The repeater station receives the signal from the mobile station on one frequency, known as the input frequency, and then retransmits the same signal on a second frequency which is known as the output frequency. The difference between these two frequencies is called the frequency offset.

In the UK the input channel is generally 600 kHz below the output frequency for repeaters working in the 2-M (145-MHz) band. On the 70-cm (433-MHz) band the repeaters use a frequency offset of 1600 kHz. Here the input channel is on the higher frequency. On the 23-cm (1300-MHz) band the frequency offset is 6 MHz, with the input signal on the lower frequency.

In North America there are hundreds of repeaters operating on the VHF and UHF bands. American repeaters use a variety of input and output frequency offsets. On 29 MHz there are FM repeaters which operate on channels in the 29 500–29 700-kHz band with an input-to-output frequency offset of 100 kHz. Inputs are in the band 29.51–29.59 MHz and outputs are in the band 29.61–29.69 MHz. There are several 29-MHz repeaters operating in North America and some are being planned for the European area.

For proper operation only one station can transmit through the repeater at a time. Users take it in turn to talk through the repeater. Access to the repeater may be controlled by carrier detection when the repeater will start relaying signals when it detects a carrier on its input. To avoid false triggering of the repeater some systems will wait until modulation is detected before opening. Access to many repeaters is achieved by sending a 500-ms-long tone burst with a frequency of 1750 Hz; the repeater opens at the end of the tone signal. Most handheld and mobile transceivers have an automatic tone burst generator which is activated when the push to talk (PTT) button is pressed.

Once the repeater has been triggered by a carrier or tone burst it will relay signals received on the input channel until the receiver detects a loss of input signal. When the input signal is dropped, the repeater usually sends a K in Morse on its output to indicate to other users that it is available for the next station to access it.

To prevent one station hogging the repeater, a time-out system is generally used so that after a period of perhaps 2 min the repeater will switch to beacon mode. In this mode the repeater transmits a tone on its output channel until the input is released, or it receives a new tone burst. If the repeater is not accessed again after it sends its K signal then it will remain dormant until a new station sends an access signal on the input channel. At intervals of about 15 min the repeater will transmit its callsign in Morse to identify itself.

Two-metre repeaters in the UK

On the 2-m band there are eight repeater channels available. Several repeaters are allocated to each channel but are located so that under normal propagation conditions they do not interfere with one another. The channels are numbered R0 to R8 and the inputs occupy the channels that would logically have been simplex channels S0 to S8.

Channel R0: input 145.000 MHz, output 145.600 MHz

GB3AS	Caldbeck Cumbria
GB3CF	Leicester, Leicestershire
GB3EL	East London
GB3FF	Fife, Scotland
GB3LY	Limavady, Northern Ireland
GB3MB	Bury, Lancashire
GB3SR	Brighton, East Sussex
GB3SS	Elgin, Scotland
GB3YC	Scarborough, Yorkshire

Channel R1: input 145.024 MHz, output 145.625 MHz

GB3GD	Snaefell, Isle of Man
GB3HC	North Yorkshire
GB3KS	Dover, Kent
GB3MH	Malvern Hills, Worcestershire
GB3NB	Wymondham, Norfolk
GN3NG	Grampian, Scotland
GB3PA	Paisley, Scotland
GB3SC	Bournemouth, Dorset
GB3SI	St Ives, Cornwall
GB3WL	West London

Channel R2: input 145.050 MHz, output 145.650 MHz
GB3HS   Hull, Humberside
GB3MN   Stockport, Cheshire
GB3OC   Orkney Islands, Scotland
GB3PO   Ipswich, Suffolk
GB3SB   Selkirk, Scotland
GB3SL   South London
GB3TR   Torquay, Devon
GB3WH   Swindon, Wiltshire

Channel R3: input 145.075 MHz, output 145.675 MHz
GB3BM   Dudley, West Midlands
GB3DR   Dorset
GB3ES   Hastings, East Sussex
GB3LC   Argyll, Scotland
GB3LU   Lerwick, Scotland
GB3NA   Barnsley, Yorkshire
GB3PE   Peterborough, Cambridgeshire
GB3PR   Perth, Scotland
GB3RD   Reading, Berkshire
GB3SA   Swansea, Wales
GB3TE   St Othys, Essex

Channel R4: input 145.100 MHz, output 145.700 MHz
GB3AR   Carmarfon, Wales
GB3BB   Brecon, Wales
GB3BT   Berwick on Tweed
GB3EV   Appleby, Cumbria
GB3HH   Buxton, Derbyshire
GB3HI   Isle of Mull, Scotland
GB3KN   Maidstone, Kent
GB3VA   Aylesbury, Buckinghamshire
GB3WD   Princeton, Devon

Channel R5: input 145.125 MHz, output 145.725 MHz
GB3BI   Inverness, Scotland
GB3DA   Danbury, Essex
GB3LM   Lincoln, Lincolnshire
GB3NC   St Austell, Cornwall
GB3NI   Belfast, Northern Ireland
GB3SN   Fourmarks, Hampshire
GB3TP   Keithley, West Yorkshire
GB3TW   Burnhope, Durham
GB3VT   Stoke on Trent, Staffordshire

Channel R6: input 145.150 MHz, output 145.750 MHz

GB3AM	Birmingham, West Midlands
GB3BC	Newport, Wales
GB3CS	Black Hill, Scotland
GB3MP	Moel-y-Parc, Wales
GB3PI	Barkway, Hertfordshire
GB3TY	Hexham, Northumberland
GB3WS	Horsham, West Sussex

Channel R7: input 145.175 MHz, output 145.775 MHz

GB3DG	Dumfries, Scotland
GB3FR	Old Bolingbroke, Lincolnshire
GB3GN	Aberdeen, Scotland
GB31G	Stornaway, Scotland
GB3NL	North London
GB3PC	Portsmouth, Hampshire
GB3PW	Powys, Wales
GB3RF	Burnley, Lancashire
GB3WK	Leamington Spa
GB3WT	West Tyrone, Northern Ireland
GB3WW	Dyfed, Wales

Note that from time to time existing repeaters may change their location or frequency and new repeaters are added to the network.

432-MHz repeaters in the UK

The 432-MHz band contains 16 repeater channels. Several repeaters are allocated to each channel but are located to avoid interference under normal conditions. These repeaters are numbered RB0 to RB15 and their outputs occupy the frequencies that would logically have been simplex channels SU0 to SU15.

Channel RB0: input 434.600 MHz, output 433.000 MHz

GB3BN	Bracknell, Berkshire
GB3CK	Ashford, Kent
GB3DT	Blandford Forum, Dorset
GB3EX	Exeter, Devon
GB3LL	Llandudno, Wales
GB3MK	Milton Keynes, Buckinghamshire
GB3MS	Malvern Hills, Worcestershire
GB3NR	Norwich, Norfolk
GB3NT	Gateshead
GB3NY	Scarborough, North Yorkshire
GB3PF	Blackburn, Lancashire

GB3PU   Perth, Scotland
GB3SO   Boston, Lincolnshire
GB3SV   Bishops Stortford, Hertfordshire
GB3US   Sheffield
GB3WN   Wolverhampton

Channel RB1: input 434.625 MHz, output 433.025 MHz
GB3BA   Aberdeen, Scotland
GB3BV   Hemel Hempstead, Hertfordshire
GB3DV   Doncaster
GB3HO   Horsham, West Sussex
GB3MA   Bury, Lancashire

Channel RB2: input 434.650 MHz, output 433.050 MHz
GB3AV   Aylesbury, Buckinghamshire
GB3CH   Plymouth, Devon
GB3CI   Corby, Northamptonshire
GB3EK   Margate, Kent
GB3FC   Fylde Coast, Lancashire
GB3LS   Lincoln, Lincolnshire
GB3LV   North London
GB3NN   Wells, Norfolk
GB3NX   Crawley, West Sussex
GB3OS   Stourbridge, Worcestershire
GB3PH   Portsmouth, Hampshire
GB3ST   Stock on Trent, Staffordshire
GB3UL   Belfast, Northern Ireland
GB3YS   Yeovil, Somerset

Channel RB3: input 434.675 MHz, output 433.075 MHz
GB3CC   Chichester, West Sussex
GB3ER   Danbury, Essex
GB3HL   West London
GB3HU   Hull, Humberside
GB3MD   Mansfield, Nottinghamshire
GB3NH   Northampton, Northamptonshire
GB3VS   Taunton, Somerset

Channel RB4: input 434.700 MHz, output 433.100 MHz
GB3AN   Anglesey, Wales
GB3GC   Goole, Humberside
GB3IH   Ipswich, Suffolk
GB1IW   Isle of Wight
GB3KL   Kings Lynn, Norfolk
GB3KN   Elgin, Scotland
GB3KR   Kidderminster, Worcestershire

GB3LE   Leicester, Leicestershire
GB3NK   Wrotham, Kent
GB3OH   West Lothian, Scotland
GB3SP   Pembroke, Wales
GB3UB   Bath, Avon

Channel RB5: input 434.725 MHz, output 433.125 MHz
GB3EB   Brentwood, Essex
GB3GH   Gloucester, Gloucestershire
GB3HY   Haywards Heath, West Sussex
GB3IM   Isle of Mull, Scotland
GB3NW   North West London
GB3OV   Huntingdon
GB3WB   Weston Super Mare
GB3WJ   Scunthorpe

Channel RB6: input 434.750 MHz, output 433.150 MHz
GB3BD   Bedford, Bedfordshire
GB3BR   Brighton, Sussex
GB3CD   Derby, Derbyshire
GB3CR   Mold, Wales
GB3CW   Powys, Wales
GB3HB   St Austell, Cornwall
GB3HC   Hereford
GB3LW   Central London
GB3ME   Rugby, Warwickshire
GB3SK   Canterbury, Kent
GB3SY   Barnsley, South Yorkshire
GB3WC   Port Talbot, Wales

Channel RB7: input 434.775 MHz, output 433.175 MHz
GB3BL   Bedford, Bedfordshire
GB3HZ   High Wycombe, Buckinghamshire
GB3MF   Macclesfield, Cheshire
GB3NM   Mapperley, Nottinghamshire
GB3WY   Queensbury, West Yorkshire

Channel RB8: input 434.800 MHz, output 433.200 MHz
GB3CM   Carmarthen, Wales
GB3EH   Edge Hill, Warwickshire
GB3LA   Leeds
GB3PY   Madingley, Cambridgeshire
GB3TF   Telford

Channel RB9: input 434.825 MHz, output 433.225 MHz
GB3BE   Bury St Edmunds
GB3CL   St Osyth, Essex

GB3CV   Coventry, West Midlands
GB3HD   Huddersfield, Yorkshire
GB3RC   Nantwich, Cheshire
GB3SW   Salisbury, Wiltshire

Channel RB10: input 434.850 MHz, output 433.250 MHz
GB3DD   Dundee, Scotland
GB3DY   Wirksworth, Derbyshire
GB3LI   Liverpool
GB3LT   Luton, Bedfordshire
GB3ML   Black Hill, Scotland
GB3MW   Leamington Spa
GB3NS   Banstead, Surrey
GB3PB   Peterborough
GB3PO   Grampian, Scotland

Channel RB11: input 434.875 MHz, output 433.275 MHz
GB3AH   Swaffham, Norfolk
GB3BK   Reading, Berkshire
GB3DC   Sunderland
GB3GR   Grantham, Lincolnshire
GB3GY   Grimsby, Humberside
GB3HN   Hitchin, Hertfordshire
GB3HT   Hinckley, Leicestershire
GB3LR   Newhaven, East Sussex
GB3NF   Fawley, Hampshire
GB3RE   Chatham, Kent
GB3SH   Honiton, Devon
GB3WP   Manchester

Channel RB12: input 434.900 MHz, output 433.300 MHz
GB3EE   Chesterfield
GB3GF   Guildford, Surrey
GB3MT   Bolton, Lancashire
GB3PT   Barkway, Hertfordshire

Channel RB13: input 434.925 MHz, output 433.325 MHz
GB3CA   Carlisle, Cumbria
GB3CY   York, Yorkshire
GB3DS   Worksop, Nottinghamshire
GB3GU   Guernsey, Channel Islands
GB3HW   Romford, Essex
GB3LC   Louth, Lincolnshire
GB3SM   Leek, Staffordshire
GB3TD   Swindon, Wiltshire

GB3VH   Welwyn Garden City, Hertfordshire
GB3XX   Daventry, Northamptonshire

Channel RB14: input 434.950 MHz, output 433.350 MHz
GB3AB   Aberdeen, Scotland
GB3CB   Birmingham, West Midlands
GB3CC   Blaenau Ffestiniog, Wales
GB3CE   Colchester, Essex
GB3ED   Edinburgh, Scotland
GB3GL   Glasgow, Scotland
GB3HE   Hastings, East Sussex
GB3HK   Hawick, Scotland
GB3HR   Stanmore, Middlesex
GB3LF   Lancaster
GB3MR   Stockport
GB3ND   Great Torrington, Devon
GB3SD   Weymouth, Dorset
GB3TL   Spalding, Lincolnshire
GB3TS   Middlesborough, Cleveland
GB3WF   Near Leeds
GB3YL   Lowestoft, Suffolk

Channel RB15: input 434.975 MHz, output 433.375 MHz
GB3FN   Farnham, Surrey
GB3LH   Near Aylesbury
GB3OM   Omagh, Northern Ireland
GB3OX   Oxford, Oxfordshire
GB3PP   Preston, Lancashire
GB3SG   Cardiff, Wales
GB3SU   Sudbury, Suffolk
GB3SZ   Bournemouth, Dorset
GB3TH   Tamworth, Staffordshire
GB3WI   Wisbech, Cambridgeshire
GB3WU   Wakefield, Yorkshire

1.3-GHz repeater channels in the UK

On this band there are 20 repeater channels but only a few of these are currently occupied by operational repeaters.

Channel RM0: input 1291.000 MHz, output 1297.000 MHz
GB3GH   Bushey Heath, Hertfordshire
GB3MC   Bolton, Lancashire
GB3NO   Norwich, Norfolk

Channel RM1: input 1291.025 MHz, output 1297.025 MHz

Channel RM2: input 1291.050 MHz, output 1297.050 MHz

Channel RM3: input 1291.075 MHz, output 1297.075 MHz
GB3CP   Crawley, West Sussex
GB3PS   Barkway, Hertfordshire
GB3SE   Stoke on Trend, Staffordshire

Channel RM4: input 1291.100 MHz, output 1297.100 MHz

Channel RM5: input 1291.125 MHz, output 1297.125 MHz

Channel RM6: input 1291.150 MHz, output 1297.150 MHz
GB3AD   Southampton, Hampshire
GB3BW   Bedford, Bedfordshire
GB3MM   Wolverhampton, West Midlands

Channel RM7: input 1291.175 MHz, output 1297.175 MHz

Channel RM8: input 1291.200 MHz, output 1297.200 MHz

Channel RM9: input 1291.225 MHz, output 1297.225 MHz
GB3RU   Reading, Berkshire
GB3WX   Brighton, Sussex

Channel RM10: input 1291.250 MHz, output 1297.250 MHz

Channel RM11: input 1291.2750 MHz, output 1297.275 MHz

Channel RM12: input 1291.300 MHz, output 1297.300 MHz

Channel RM13: input 1291.325 MHz, output 1297.325 MHz

Channel RM14: input 1291.350 MHz, output 1297.350 MHz

Channel RM15: input 1291.375 MHz, output 1297.375 MHz
GB3LN   North London

Channel RM16: input 1291.400 MHz, output 1297.400 MHz

Channel RM17: input 1291.425 MHz, output 1297.4250 MHz

Channel RM18: input 1291.450 MHz, output 1297.450 MHz

Channel RM19: input 1291.475 MHz, output 1297.475 MHz

From time to time new repeaters are activated and some existing repeaters may change location or frequency. A full list of the current repeaters together with their locations is available from the RSGB.

Linear transponders

One problem with the conventional repeater is that only one contact can take place via the repeater at any time. By using an alternative device known as a linear transponder, several amateur contacts can be made simultaneously. In the transponder a wideband receiver is used at the input and the received signals are mixed with a local oscillator and translated to a new band of frequencies which are then amplified and transmitted.

A typical transponder might operate with a 100-kHz wideband of input signals so that several stations on different frequencies within the band may operate simultaneously. This follows the same practice as that used for transponders on an amateur space satellite.

Input and output signals on a linear transponder are usually either CW or SSB, and any number of contacts may take place simultaneously via the transponder. The transmitter power is shared among the various signals in the band in proportion to the received signals. Narrow-band FM signals could be used but they take up about ten times the bandwidth of an SSB signal so that the number of users would be severely limited. RTTY or packet transmissions, however, could be used quite effectively on a transponder.

Currently two 100-kHz segments of the 1.3-GHz band are set aside for use by transponders, with one segment (1296.5–1296.6 MHz) used as an input band and the other (1296.6–1296.7 MHz) as an output band.

Emergency operation (RAYNET)

The Radio Amateur Emergency Network (RAYNET) has been set up in the UK to provide assistance to the emergency services and the police when an emergency such as a flood or other disaster occurs. The country is divided up into 12 regions, each of which has a RAYNET regional organizer and a number of area organizers. In the event of an emergency occurring, the members of the local network may be alerted and the amateurs then provide assistance with communications using amateur radio channels. Most of this communication is carried out on the VHF bands but for some international disasters, such as earthquakes, the HF bands may be used to provide contacts with the emergency area.

At regular intervals the RAYNET groups may take part in practice exercises in preparation for any real emergency that might occur. Apart from emergency use, the RAYNET communications facilities may also be used at sports events such as marathon runs to provide communications for first aid requests or control of the event.

In the USA similar schemes have been set up by the radio amateurs. These are the Amateur Radio Emergency Service (ARES), organized by ARRL, and the Radio Amateur Civil Emergency Service (RACES), which is part of the civil defence organization. Like RAYNET, these amateur emergency services work in association with official services to provide communications assistance during emergencies or disasters.

In the UK a number of frequencies on the VHF and UHF bands have been set aside for use by RAYNET during emergencies and events such as marathons and cycle races where RAYNET provides communications support.

Frequency (MHz)	Mode
70.350	FM
70.375	FM
70.400	FM
144.260	SSB
144.775	FM
144.825	FM
145.200	FM (S8)
145.225	FM (S9)
145.800	FM
433.700	FM (SU25)
433.725	FM (SU26)
433.750	FM (SU27)
433.775	FM (SU28)

The CAIRO interconnect system

One problem that is encountered by amateurs when setting up a multi-operator station for contest or RAYNET operation is that of incompatibility between the connectors for the various items of equipment brought along by members of the group. This often means that before the station can be set up a number of interconnecting leads have to be quickly patched together to link the various pieces of equipment. In an attempt to overcome this problem the CAIRO interconnect system was devised by the Electronics Engineering Department at Aston University in Birmingham.

CAIRO provides a standardized interconnect system using 3-, 5- or 7-pin DIN connectors which carry the microphone, PTT and audio output signals between, say, a local operating position and a remote transceiver. The scheme makes use of the fact that a 3-pin or 5-pin plug will mate with a 7-pin DIN socket and the pins will retain the same number sequence on the two connectors. The CAIRO scheme is arranged so that the signals from the transceiver are always presented on a 7-pin DIN socket. Remote accessories use 3-, 5- or 7-pin plugs as appropriate.

The CAIRO 7 connection scheme on a 7-way connector is:

Pin	Signal
1	Primary audio
2	Screens + ground
3	Microphone hi (+)
4	PTT (push to talk)
5	Microphone lo (−)
6	Secondary audio
7	DC power + 12 V, 1 A

The primary audio is the normal external speaker output of the transceiver. The PTT is operated by making a contact between pin 4 and ground. Secondary audio is 1-V peak fixed level audio, which is a copy of the audio on pin 1. The microphone hi line (pin 3) may also carry a superimposed DC voltage of about 5 V for energizing electret microphones.

By using a distribution box with several 7-pin socket outlets a number of separate operators may access a single transceiver rather than using individual transceivers. For a remote loudspeaker a 3-pin DIN plug may be inserted into the 7-pin distribution socket. For normal transceiver operation a 5-pin DIN connector may be used at the remote end, giving microphone and PTT control as well as output audio. The full 7-pin connection scheme provides an additional audio output line and a + 12 V supply for powering equipment at the remote operating position.

The cable used for normal remote operations is a four-way individually screened type connected to pins 1 to 5 of the connector. Two separate cores are used for the microphone so that the earth connection for this circuit is made only at the transceiver end. The cable screens are all connected to pin 1 of the connector. Pin 3 of the connector may also carry a superimposed DC voltage of about 5 V, which is used for powering electret microphones at the operator end of the system.

For long cable a reel may be used and this would be sited at the operating position. The free end would have a 7-pin plug to connect to the transceiver output lead, and the reel might have several 7-pin sockets connected in parallel to allow various operating accessories, such as microphones, headsets and Morse keys, to be connected to the system. These accessories would have leads with 3-, 5- or 7-pin DIN plugs fitted as appropriate.

CAIRO 8

An extension to the basic 7-pin CAIRO scheme has been devised for handling the signals needed for remote operation with packet terminal node controllers (TNCs) or RTTY/AMTOR equipment. These units may require a Squelch signal to indicate that squelch has been broken by the incoming signal and may derive their power supply from the remote power line on pin 7 of the CAIRO connector.

The connections for CAIRO 8 are:

Pin	Signal
1	Primary audio
2	Ground and cable screen
3	Microphone (hi (+)
4	PTT
5	Microphone lo (−)
6	Audio line 1 V peak
7	DC power + 12 V at 1 A
8	Squelch

The cable used for the full CAIRO 8 line is an eight-core overall screened type with two cores used for the 12-V DC supply feed. Note that the CAIRO 8 scheme is fully compatible with the basic CAIRO system, since the 8-pin socket can accept 3-, 5- or 7-pin plugs.

Further details of the CAIRO interconnect system can be obtained from:

The CAIRO Laboratory
Department of Electronic Engineering and Applied Physics
Aston University
Aston Triangle
Birmingham B4 7ET
UK

Amateur radio rallies and conventions

During the year radio amateurs get together at a number of radio rallies and conventions. Rallies were initially developed as meetings of amateur mobile operators and these rallies usually feature trade and amateur radio stalls selling new or secondhand equipment and components. Conventions feature a series of lectures or demonstrations on a range of subjects of interest to radio amateurs. Most conventions also have trade and club stalls. Some of the larger rallies and conventions in the UK and America are as follows:

February	Great Northern Rally, Manchester
March	London Amateur and Computer Show, London
March/April	RSGB National Convention, Birmingham
March/April	RSGB VHF Convention, Esher, Surrey
April	Dayton Hamvention, Dayton, Ohio, USA
July	Longleat Rally, near Warminster, Wiltshire
August	National Rally, Woburn Abbey, Bedfordshire

There are many other rallies and events throughout the year and details of these are usually published in the amateur radio magazines.

Jamboree on the Air

Amateur radio has been associated with the Scouting movement world-wide, and every year for one weekend during October radio amateurs set up stations at scout camps and meeting halls to hold a jamboree on the air. In many countries special permission is given to allow the scouts themselves to pass messages and greetings to other scout groups around the world. The radio jamboree also allows those scouts taking tests in communications to work towards obtaining their badges.

In the UK many of the stations taking part in the jamboree on the Air operate with special GB call signs which are allocated for the period covering the jamboree week.

A similar series of special event stations take to the air in association with the Guides during the week of their Thinking Day.

5

Radioteletype (RTTY)

The most familiar means of communication by radio is telephony, in which speech signals are transmitted using analogue techniques. There are, however, various alternative communication schemes in which the information is conveyed using digital codes. These schemes are generally referred to as telegraphy.

The simplest form of telegraphy uses the Morse code, which is generated by switching the transmitter output on and off to produce a sequence of short and long pulses known as dots and dashes. In most amateur stations the Morse code symbols are generated manually by using a Morse key to switch the transmitter. This mode of transmission is referred to as manual telegraphy.

Another approach to sending messages is radioteletype (RTTY), which is based on the automatic telegraphy systems originally devised for sending telegrams and other messages. In the Morse code system each letter may consist of a varying number of dots and dashes. The most frequently used letters, such as E and T, have just one dot or dash, whilst numbers consist of a group of five dot and dash units. Although this is well suited for manual decoding, the varying length of the symbol code makes automatic decoding more difficult to implement. For automatic telegraphy systems it is more convenient to use a fixed-length code for each character that is to be sent.

In the early RTTY system each letter code consisted of a sequence of seven pulses, of which five are used to define the actual letter or number. The codes were generated by a machine with a keyboard similar to that of a typewriter, and as each key was pressed the appropriate character code was transmitted. Often the message would be typed in before transmission and the teleprinter would produce a perforated paper tape where a pattern of holes was punched across the tape for each letter typed in. To transmit the message the punched paper tape was passed through a tape reader, which then generated the codes automatically at a constant speed. This approach allowed the message to be sent at a much faster rate than the operator could type it in. The system also had to operate directly so that as the operator typed each letter the appropriate code was generated and sent out.

At the receiving end the received pulse groups were fed to a decoder unit, which operated a printer to provide a printed record of the received text. This basic RTTY transmission scheme is widely used today by both commercial and amateur stations. In modern systems a computer-based system may be used at both ends of the link in place of the old teleprinter machines, and printout is produced by a modern dot matrix printer. In recent years various developments of RTTY, such as the SITOR, forward error control (FEC) and automatic repeat request (ARQ) systems, have been introduced. These new systems include techniques which can detect and correct errors when reception conditions are not ideal.

Baudot code

Amateur RTTY is based upon the early commercial automatic radio telegraphy systems. The data for the text symbols are sent using an asynchronous serial code system in which each letter or symbol is represented by a serial sequence of five pulses or bits. Each bit may have one or two states, which are known as 'mark' and 'space'. When no message is being sent, the signal is held in the mark state. The sequence of data bits for sending a single character is shown in Figure 5.1.

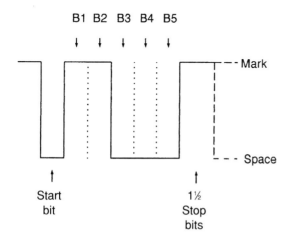

Figure 5.1 Sequence of data bits for a single RTTY character using a 5-bit Baudot code

| LTRS | G | FIGS | 3 | LTRS | F | Z | X |

Figure 5.2 Use of LTRS and FIGS shift codes when sending a callsign via Baudot RTTY

Before the character code sequence begins, the signal is at the mark level. To indicate to the receiver that a new character code is about to be sent, a single start bit at the space level is transmitted. This start bit provides a timing reference which allows the receiver and transmitter circuits to be synchronized. At the receiving end the detection of the start bit starts a clock which defines the timing points for detecting the states of the following character data bits.

The five bits representing the character are then sent in sequence after the start bit, with the least significant bit (bit 1) being transmitted first. At the receiving end the signal is sampled roughly at the middle of each data bit period and the states of the five bits are then stored in a register. Each of the possible combinations of the states of the five data bits causes one particular character to be printed or displayed.

To complete the code sequence the signal returns to the mark level for 1.5 bit periods to produce a stop signal. The stop signal ensures that successive character codes are separated by a period at the mark level so that the receiver can always detect the start bit of the next character code to maintain correct synchronization. Since the system sends one character code at a time and there may be varying gaps between characters, depending upon the operator's typing ability, this mode of transmission is called asynchronous start–stop telegraphy.

The code normally used for RTTY transmissions is the ITU No. 2 alphabet, which is often referred to as the Baudot code. Each data bit is allocated a fixed time slot. If the bit is to be a 1, the signal is at the mark level, whilst for a 0 state the signal is at the space level during the time slot.

The 5-bit data code allows 32 possible combinations which are not sufficient to handle both letters and figures. To overcome this limitation a shift system is used to switch between letters and figures. When figures are to be sent, a special FIGS shift code is transmitted. When this code is detected, the receiver decoder switches to the figures mode and received codes are printed as figures or special signs. To return to the normal letters mode the LTRS shift code is sent (Figure 5.2).

In the letters mode only capital letters are available. The control codes such as line feed, carriage return, space and the FIGS and LETRS shift codes are common to both the letters and figures modes.

ITU No. 2 code for RTTY (Baudot)

Code bits 54321	Hex value	Letters	Figures
00011	03	A	–
11001	19	B	?
01110	0E	C	:
01001	09	D	$
00001	01	E	3
01101	0D	F	!
11010	1A	G	&
10100	14	H	
00110	06	I	8
01011	0B	J	Bell
01111	0F	K	(
10010	12	L	)
11100	1c	M	.
01100	0C	N	,
11000	18	O	9
10110	16	P	0
10111	17	Q	1
01010	0A	R	4
00101	05	S	,
10000	10	T	5
00111	07	U	7
11110	1E	V	=
10011	13	W	2
11101	1D	X	/
10101	15	Y	6
10001	11	Z	+
01000	08	CR (carriage return)	
00010	02	LF (line feed)	
11111	1F	LTRS (letter shift)	
11011	1B	FIGS (figure shift)	
00100	04	SP (space)	
00000	00	BLK (blank)	

Some RTTY printers or displays designed for the US market may have slightly different codes for some of the symbols in the FIGS mode, but the character set shown will be correct for most transmissions.

Some countries use special codes which have an extra shift code to allow more characters to be accommodated. Examples are the Cyrillic code used by some Russian stations and the Arabic code. These codes will produce what appears to be gibberish on a standard RTTY decoder system, and special decoders are required to produce the correct output.

Control sequences

Various symbol sequences are used by commercial telegraph operators to signify control actions. Some of these are also used by radio amateurs. Some popular codes are:

ZCZC	Start of message
NNNN	End of message
XXXXX	Error

SITOR and AMTOR

In the simple RTTY system there is no protection against errors in the message when the received data become corrupted by interference or by signal fading. To overcome this problem various new coding systems have been devised which allow the detection and correction of errors in the received message.

One such system is SITOR, which was developed for use in the commercial maritime radio service. There are two modes of operation in the SITOR system which are known as ARQ and FEC. Both modes provide facilities for detecting and correcting errors in the received signal to give a more reliable RTTY-type communications link. AMTOR is the amateur radio version of the SITOR transmission system. A number of other transmission schemes based on ARQ and FEC principles but with different transmission rates are used by military, press and commercial stations.

The SITOR coding system uses the same basic character set and shift modes as the conventional 5-bit RTTY transmissions but each character code contains seven data bits instead of five. The 7-bit code allows up to 128 combinations but of these only 35 are actually used. The SITOR codes are chosen so that the bit pattern always contains four mark bits and three space bits. At the receiving end a simple check of the number of mark and space bits in the received code enables an error to be detected, and action can be taken to correct the received data.

In SITOR there are no start or stop bits and the blocks of seven data bits for each character code are sent one after another at a rate of 10 ms per bit

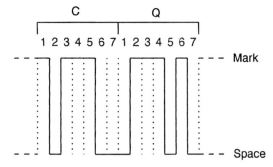

Figure 5.3 Serial data signal for sending the letters CQ using the 7-bit SITOR/AMTOR code

(100 baud). In the ARQ the characters are sent in groups of three, whilst in FEC mode they are sent in a continuous stream. This mode of operation is known as a synchronous transmission.

The codes are sent with the least significant bit first in the same way as conventional RTTY. Figure 5.3 shows the sequence of bits for sending the characters CQ. Synchronization between receiver and transmitter is achieved by using certain special idling codes when no message data are being sent. Alternatively the decoder may examine the incoming bit stream for a valid SITOR character code with four ones and three zeros.

SITOR/AMTOR data code (CCIR 476)

Code word binary	Hex value	Letters	Figures
1000111	47	A	–
1110010	72	B	?
0011101	1D	C	:
1010011	53	D	$
1010110	56	E	3
0011011	1B	F	!
0110101	35	G	&
1101001	69	H	
1001101	4D	I	8
0010111	17	J	Bell
0011110	1E	K	(
1100101	65	L	)
0111001	39	M	.
1011001	59	N	,
1110001	71	O	9
0101101	2D	P	0
0101110	2E	Q	1
1010101	55	S	'
1110100	74	T	5
1001110	4E	U	7
0111100	3C	V	–
0100111	27	W	2
0111010	3A	X	/
0101011	2B	Y	6
1100011	63	Z	+
1111000	78	CR (carriage return)	
1101100	6C	LF (line feed)	
1011010	5A	LTRS (letter shift)	
0110110	36	FIGS (figure shift)	
1011100	5C	SP (space)	
1101010	6A	BLK (blank)	
1100101	65	CS1 (control 1)	
1101010	6A	CS2 (control 2)	
1001101	4D	CS3 (control 3)	
0001111	0F	Idle (alpha)	
0110011	33	Idle (beta)	
1100110	66	Signal repetition (RQ)	

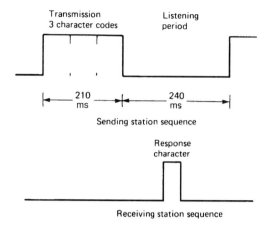

Figure 5.4 The timing sequence for send and response signals in AMTOR ARQ mode

ARQ mode

The usual mode of operation for AMTOR/SITOR stations is ARQ (automatic repeat request), which is also known as mode A. In this mode the transmitting station sends data in groups of three characters and after each group the receiving station replies with an acknowledgement signal which indicates whether the three character codes have been correctly received. The sequence is repeated every 450 ms. This sequence is shown in Figure 5.4.

The three characters occupy 70 ms each, giving a total of 210 ms of signal from the sending station. During the remaining 240 ms the receiving station sends back a single-character code indicating the status of the received signals. Normally the receive station will send the CS1 and CS2 codes alternately. If an error is detected the receive station repeats the code that was sent for the previous block. When an error is signalled, the transmitting station repeats the transmission of the last group of three character codes until reception is successfully acknowledged.

The ARQ mode is normally used for communication between two individual stations since it requires interaction between the two stations for proper operation. This type of operation is easily recognized by the regular chirping sound as the individual character groups are sent. There is also a noticeable change in the chirping sound when the station switches from sending three-symbol groups to sending just the single-character acknowledgement codes. Other stations may listen to a contact operating in this mode but the message may appear rather corrupted on the screen if there are errors and repetitions.

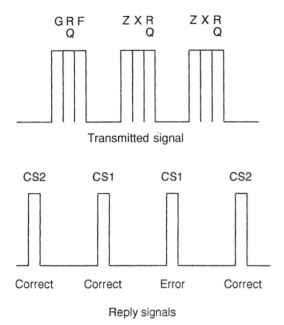

Reply signals

Figure 5.5 Examples of the ARQ mode calling sequence and repetition for error correction

An important aspect of the ARQ mode is that the transmit–receive changeover operation at each station must be rapid, as otherwise the acknowledgement signal may be missed. This mode can present problems on long-distance contacts where the signal delay over the round trip path between the stations may exceed the available time in which the acknowledgement must be received.

ARQ calling sequence
In the ARQ mode the station which starts the contact becomes the master and controls the timing sequence of the signals. The calling sequence consists of two three-character blocks which may be repeated until a contact is established. In the first block the second character is a signal repetition (RQ) and in the second block the third character is an RQ code. The other characters in the two blocks are normal character codes which may represent the call sign of the calling station. This arrangement is shown in Figure 5.5. To establish contact the receiving station responds with alternate CS1 and CS2 codes. On receipt of two successive CS1 (or CS2) codes the transmitting station may start to send messages.

SITOR/AMTOR station identifiers

In the calling sequence, four letters are available to identify the station. This arrangement fits in well with the maritime applications for which SITOR was designed, since ships use four-letter call signs.

For amateur applications the station callsign is usually longer, so an alternative identification scheme has to be used. The first letter of the amateur call and the last three letters are used. Thus G3ABC would use the code GABC. For a two-letter G call such as GGAB, the identifier becomes GGAB, since the first G is also one of the last three letters in the complete callsign. If the callsign starts with a number, this is ignored, so that 2E1XYZ would use the identifier EXYZ. The problem is that G4ABC and GM3ABC would also end up with the same GABC identifier as G3ABC. If someone else is already using the code that fits your callsign then you will need to alter your identifier by changing one of the last three letters. Remember that only a small percentage of amateurs in any given country use AMTOR, so the identifier codes can be decided by mutual agreement between stations. The general rule is that the first station to use an identifier code keeps it.

ARQ changeover sequence

To change the direction of data flow in the ARQ mode the sending station transmits the sequence (FIGS + ?). On receipt of this the receiving station sends a CS3 code as its response. When the sending station detects the CS3 code it transmits the sequence (beta alpha beta). On receipt of this sequence the receiving station changes to the transmit mode and sends a block of three RQ codes if it is a slave station. If the receiving station is a master this block contains one RQ and two other characters. When the sending station detects the RQ code it switches to receive and changeover is complete.

If the receiving station wishes to change to send mode it transmits a CS3 code in response to a received block. On receipt of the CS3 code the sending station transmits the sequence (beta alpha beta) and the changeover occurs as before.

ARQ contact termination

The sending station may terminate a contact by sending the 'end of communication' sequence (Alpha Alpha Alpha). If a receiving station wishes to terminate the contact it must first initiate a changeover so that it becomes the sending station and then it transmits the end of communication sequence (Alpha Alpha Alpha). Communication may also be terminated if continuous errors have occurred in the last 64 successive blocks of data.

FEC broadcast mode

The alternative mode of operation for AMTOR/SITOR is FEC (forward error control) or more simply mode B. In this mode the transmitting station

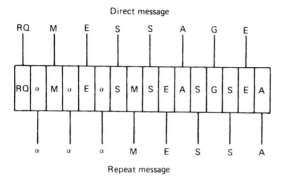

Figure 5.6 The character transmission sequence for AMTOR FEC broadcast mode

sends the message twice with the repeat version interleaved with the original message sequence. The primary message is sent using the odd-numbered symbol slots in the message stream. The repeated message is sent in the even-numbered slots but is delayed five words behind the main message. This is shown in Figure 5.6. In this mode data are sent in a continuous stream rather than in blocks of three characters.

At the start of transmission both the message slots and the repeat slots are filled with idle characters to allow the receiving station to synchronize its decoding circuits. Here the message slots contain a CS1 code and the repeat slots contain a CS2 code to allow the receiving stations to synchronize correctly. This idling pattern is also sent between messages to maintain synchronism.

The advantage of this mode is that it does not require an acknowledgement signal from the receiving station and it may be correctly decoded by several stations. This mode is suited for broadcast transmissions, such as CQ calls, which are intended for reception by a number of receiving stations. The FEC broadcast mode is generally used for amateur news transmissions and propagation forecasts. In amateur contacts the FEC mode may be used for the initial call and when contact has been established the two stations generally switch to the ARQ mode. The FEC mode has the advantage that is is not affected by transmit–receive changeover delays or propagation delays between two distant stations, which can present problems when the ARQ mode is used.

FEC selective calling mode

There is an alternative mode of operation for FEC known as the selective calling mode. This is used in the maritime mobile service for messages intended for a specific station. In this selective mode the initial phasing sequence is sent normally but the identifier code of the receiving station and the following messages are transmitted with the data signals inverted so that each code contains three mark bits and four space bits. On receipt of its callsign the receiving station automatically inverts the received data to reproduce the correct message information. During the time between messages the 'idle beta' signal is sent. This mode is not normally used by amateur stations.

FEC error detection

At the receiving end both the initial and repeated codes are checked for possible errors. If the initial code is correct, it is printed. If an error is detected, the repeat code is checked and if correct is printed. If both codes contain an error, a space is printed. The repetition sequence effectively provides time diversity and, in general, if the code in one copy of the message is affected by interference or fading, the code in the repeated message is likely to be received correctly so that under most conditions the message is printed correctly at the receiving end.

ASCII Code RTTY

Some amateur stations use the 8-bit ASCII (American Standard Code for Information Interchange) code when transmitting RTTY. This code is widely used to represent text in home computer systems and for communications between computers. This code may also be used by amateur stations when sending computer programs to one another. The international version of the ASCII code is known as ISO7 and includes some variations to allow special symbols, such as the # sign, to be included for other countries.

Unlike the ITU No. 2 code, ASCII provides seven bits for the character code. This allows up to 128 different combinations for upper and lower case letters, numbers, and a range of other symbols without the need for shift

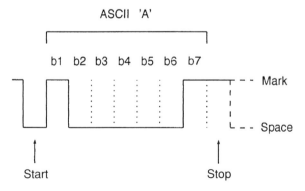

Figure 5.7 The serial data signal for a letter 'A' in ASCII code RTTY

codes. The eighth bit in the transmitted code is called the parity bit and is used for error checking.

The transmission system is asynchronous, with start and stop bits like the Baudot scheme. The code sequence for a single character is shown in Figure 5.7. Normally there will be one start bit and one stop bit, giving a total of 10 transmitted bits per character.

The parity bit may be set at either 1 or 0 to make the total number of 1 bits in the character code an odd number. This is known as odd parity coding. At the receiving end the number of 1 bits is checked: if it is odd, then the received character is correct; otherwise, an error has occurred and some action may be taken to deal with this if desired. Alternatively the parity bit may also be arranged to produce an even number of 1 bits in the code and this is known as even parity coding.

The first 32 codes in the ASCII set are reserved as control codes providing such functions as line feed, carriage return, bell etc.

ASCII code table

Decimal	Hex	Binary	Symbol
0	00	0000000	NUL
1	01	0000001	SOH
2	02	0000010	STX
3	03	0000011	ETX
4	04	0000100	EOT
5	05	0000101	ENO
6	06	0000110	ACK
7	07	0000111	BRL
8	08	0001000	BS
9	09	0001001	HT
10	0A	0001010	LF
11	0B	0001011	VT
12	0C	0001100	FF
13	0D	0001101	CR
14	0E	0001110	SO
15	0F	0001111	SI
16	10	0010000	DLE
17	11	0010001	DC1
18	12	0010010	DC2
19	13	0010011	DC3
20	14	0010100	DC4
21	15	0010101	NAK
22	16	0010110	SYN
23	17	0010111	ETB
24	18	0011000	CAN
25	19	0011001	EM
26	1A	0011010	SUB
27	1B	0011011	ESC
28	1C	0011100	FS
29	1D	0011101	GS
30	1E	0011110	RS
31	1F	0011111	US

Decimal	Hex	Binary	Symbol
32	20	0100000	Space
33	21	0100001	!
34	22	0100010	"
35	23	0100011	
36	24	0100100	$
37	25	0100101	%
38	26	0100110	&
39	27	0100111	'
40	28	0101000	(
41	29	0101001	)
42	2A	0101010	*
43	2B	0101011	+
44	2C	0101100	,
45	2D	0101101	-
46	2E	0101110	.
47	2F	0101111	/
48	30	0110000	0
49	31	0110001	1
50	32	0110010	2
51	33	0110011	3
52	34	0110100	4
53	35	0110101	5
54	36	0110110	6
55	37	0110111	7
56	38	0111000	8
57	39	0111001	9
58	3A	0111010	:
59	3B	0111011	;
60	3C	0111100	<
61	3D	0111101	=
62	3E	0111110	>
63	3F	0111111	?
64	40	1000000	@
65	41	1000001	A
66	42	1000010	B
67	43	1000011	C
68	44	1000100	D
69	45	1000101	E
70	46	1000110	F
71	47	1000111	G
72	48	1001000	H
73	49	1001001	I
74	4A	1001010	J
75	4B	1001011	K
76	4C	1001100	L
77	4D	1001101	M
78	4E	1001110	N
79	4F	1001111	O
80	50	1010000	P
81	51	1010001	Q
82	52	1010010	R
83	53	1010011	S
84	54	1010100	T
85	55	1010101	U
86	56	1010110	V
87	57	1010111	W

Decimal	Hex	Binary	Symbol
88	58	1011000	X
89	59	1011001	Y
90	5A	1011010	Z
91	5B	1011011	[
92	5C	1011100	\
93	5D	1011101	]
94	5E	1011110	^
95	5F	1011111	_
96	60	1100000	`
97	61	1100001	a
98	62	1100010	b
99	63	1100011	c
100	64	1100100	d
101	65	1100101	e
102	66	1100110	f
103	67	1100111	g
104	68	1101000	h
105	69	1101001	i
106	6A	1101010	j
107	6B	1101011	k
108	6C	1101100	l
109	6D	1101101	m
110	6E	1101110	n
111	6F	1101111	o
112	70	1110000	p
113	71	1110001	q
114	72	1110010	r
115	73	1110011	s
116	74	1110100	t
117	75	1110101	u
118	76	1110110	v
119	77	1110111	w
120	78	1111000	x
121	79	1111001	y
122	7A	1111010	z
123	7B	1111011	{
124	7C	1111100		
125	7D	1111101	}
126	7E	1111110	~
127	7F	1111111	
DEL			

ASCII control code functions

ACK	Acknowledge
	Block received OK.
BS	Backspace
	Go back one character space.
CAN	Cancel
	Cancel line (i.e. all back to last CR).
CR	Carriage return
	Return to start of line.
DC1–4	Device control codes

DEL   Delete (erase previous symbol).
DLE   Data link escape
            Changes meaning of following two or three codes.
EM   End of medium
ENQ   Enquiry
            Requests response from remote station.
EOT   End of transmission
ESC   Escape
            Selects alternative meaning for the following symbol codes.
ETB   End of transmission block
            Indicates end of a block of data or text.
ETX   End of text
            Terminates text.
FF   Form feed (new page)
FS   File separator
GS   Group separator
HT   Horizontal tab
LF   Line feed
            Move down one line.
NAK   Negative acknowledge
            Used to indicate an error in reception.
NUL   Null
RS   Record separator
SI   Shift in
            Select normal character set.
SO   Shift out
            Select alternative character set.
SOH   Start of header
            Indicates start of header section of frame.
STX   Start of text
            Indicates start of message section of frame.
SUB   Substitution
            Replace previous character.
SYN   Synchronization idle pattern
US   Unit separator
VT   Vertical tab

Baud rates

The rate of transmission of digital signals is normally measured in a unit called the baud. For simple two-state transmissions, such as RTTY, the data rate in bauds is the number of bits transmitted per second. Thus a transmission in which bits are sent at a rate of 50 per second would be referred to as a 50-baud transmission.

In an asynchronous transmission, such as standard RTTY, the baud rate is determined by the rate at which bits are sent when a character code is transmitted. It is calculated by dividing the time period for one bit into a period of 1 second. The baud rate does not give the overall rate at which data are being sent, since there may be periods between character codes where no data are actually being sent and the signal is simply held at the mark level.

For amateur transmissions the data rate for standard RTTY is usually 45.5, 50 or 75 baud. Commercial RTTY stations generally use 50 or 75 baud. When ASCII coded signals are sent, the baud rate may be increased to 110 baud on the HF bands. SITOR and AMTOR transmissions always operate at 100 baud.

The time period for each data bit for the various transmission speeds is as follows:

Baud rate	Pulse width (ms)
45	22
50	20
75	13.3
100	10
110	9
150	6.7
300	3.3
1200	0.8
2400	0.4

RTTY transmission techniques

Although the digital codes consist simply of on–off pulse signals, the RTTY data are usually transmitted by using FM techniques. On the HF bands the signal uses FSK, where the mark signal shifts the carrier perhaps 200 Hz higher than the nominal carrier frequency and space shifts it 200 Hz lower. Thus the transmitter radiates a constant power carrier and simply shifts the frequency in sympathy with the code signals.

The amount of frequency shift varies according to the service in which the RTTY signal is used as follows:

Shift (Hz)	Service
850	RTTY (weather, military)
425	RTTY (press, commercial)
170	RTTY (amateur, AMTOR and SITOR)

On the HF bands, FSK is generally used and has the emission designation F1B. Amateur stations normally use a frequency shift of 170 Hz for RTTY. The higher frequency is used to represent mark and the lower for space. The wider 425- or 850-Hz frequency shifts are used by some amateurs to give a lower error rate and easier tuning. Commercial and military stations use either 425- or 850-Hz shift for RTTY and 170 Hz for SITOR. Some commercial stations use reverse shift so that the space frequency is higher than the normal carrier frequency.

A few amateur transceivers have an FSK mode so that the data signals can be fed in directly to generate an FSK RTTY signal. In most amateur stations the FSK is achieved by switching a pair of audio tone signals to a conventional SSB transmitter where one tone represents mark and the other space. The tuning is then offset so that the mark tone produces the nominal carrier frequency +85 Hz. This type of signal is called AFSK. When this technique is used, the emission designation becomes J2B, although the resultant signal is virtually the same as FSK or F2B.

The two commonly used tones for AFSK in Europe are 1275 and 1445 Hz, which give 170-Hz shift. The higher tone is the mark frequency. For 425- and 850-Hz shifts the mark tone becomes 1700 or 2125 Hz respectively. Some other parts of the world use higher frequency tone pairs with a mark tone of 2125 Hz and space tones of 2295, 2550 or 2975 Hz. Here the space tone is the higher frequency of the pair. Most modern RTTY terminal units have a simple facility for switching the sense of the tones if required.

An important aspect to consider when operating a conventional SSB transceiver in this mode is that the power stages of amateur transceivers are often designed to handle full output power only on an intermittent basis, as would be the case for Morse or telephony. When FSK, or AFSK, RTTY signals are generated the full power is output continuously and for most transceivers the output power level should be reduced to about 50% of the maximum rating of the transmitter to avoid overheating. Some of the latest transceivers are designed with larger power amplifier heatsinks and cooling fans to permit continous full-power output for use in the FSK mode.

On VHF and UHF, RTTY transmissions may use either FSK or AFSK modulation. In the case of AFSK the two audio tones are used to frequency modulate the transmitter carrier to give a narrow-band FM signal. The frequency shift used is generally either 170 or 425 Hz, with the space tone set at 1275 Hz and the mark tone at either 1445 or 1700 Hz. The emission designation for this type of AFSK signal is F2B.

In the early days of RTTY, amateurs generally used surplus commercial teleprinters for RTTY operation, but today the personal computer has taken over this task. There are several possible set-ups which may be used.

The simplest approach is to buy a commercial unit which will detect the tones, decode the serial data and then produce a standard serial communications link to feed the computer. In this case the computer merely acts as a terminal to display and input text. The external unit usually contains its own microprocessor which performs any code translations needed. In the transmit mode the input via the computer keyboard is fed serially to the external unit, which then generates the appropriate audio output tones for the transmitter. The computer software will usually provide a display of the

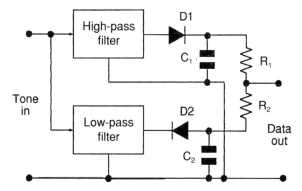

Figure 5.8 Block diagram for a filter-type RTTY tone demodulator

received text, and in a separate part of the screen the text to be transmitted, which may be typed in whilst signals are being received from the other station. One unit of this type is the AEA PK232 packet controller, which can also handle RTTY and AMTOR.

A more common approach is to use a modem to handle the tone decoding and encoding. The receiving demodulator in its simplest form consists of two narrow-bandpass filters which are tuned to the two tones representing mark and space. The audio input is fed to both filters in parallel and the outputs of the filters are rectified to produce DC signals which are fed to the inputs of an operational amplifier to produce the required logic signal output. The filters may be simple LC tuned circuits or might use a switched capacitor filter such as the MF10 integrated circuit. One advantage of the filter-type demodulator is that it is less prone to noise and interference than other demodulator techniques. This arrangement is shown in Figure 5.8.

Another popular form of demodulator is the phase-locked loop. Here an oscillator running at audio frequency is phase locked to the incoming signal. The control signal which alters the frequency of the oscillator to match it to the input signal will produce an output which corresponds to the original mark and space logic levels. This control signal is usually filtered and amplified to produce the desired logic level outputs. The signal is then fed to an input port on the computer. Many computers have a serial input port circuit which will automatically decode an asynchronous serial signal and output a parallel 5- or 8-bit code. For ASCII transmitters this gives the character code for the display, but for 5-bit baudot the signal will need to be translated before it can be displayed. AMTOR signals need to be decoded by internal software. One problem with the phase-locked loop is that it can be pulled out of lock by interference or noise and may introduce errors on weak signals. Figure 5.9 shows a typical phase-locked loop demodulator.

For receive-only operation it is possible to program most computers to perform the tone decoding and then to decode the serial signals and display

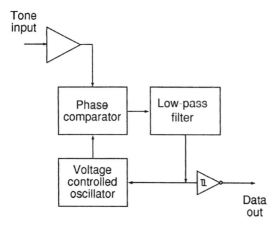

Figure 5.9 Block diagram for a phase-locked loop RTTY tone demodulator

the received text. This usually requires a simple external interface which clips the signal from the receiver to produce an audio frequency (AF) square wave. A typical circuit for this type of interface is shown in Figure 5.10. Here the signal passes through a low-pass filter to reduce high-frequency noise and is then fed to a LM 393 comparator chip. The output is a square wave of the same frequency as the input tone. This is then buffered by an emitter follower and used to feed an input line on one of the ports of the computer.

Inside the computer the tone frequency is determined by measuring the time period of each cycle of the incoming square wave. The time measurement is easiest if there is an internal timer/counter built into the computer. The counter is allowed to run continuously at a frequency about 100 times that of the incoming RTTY tones. Each time a signal transition is detected at the input

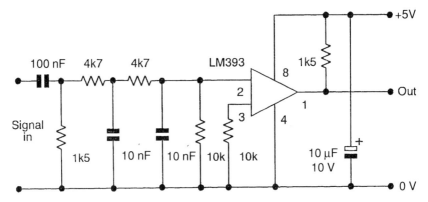

Figure 5.10 Circuit for a clipper-type computer input interface for RTTY, FAX or SSTV

port the counter reading is taken and the difference from the last counter reading is proportional to the period of the tone. From this the computer can detect which tone is present and whether the RTTY signal is at mark or space.

Once the tones have been converted into logic 0 and 1 states, the program looks for a 1 to 0 transition for the start pulse. After this there is a short delay so that the next sample of the input is taken roughly in the middle of the first data bit period. The five data bits of the word are then sampled and the data shifted through a register. At the end of this stage the character data word is in the register and the code can be used by the computer software to display the appropriate character on the screen. In the Baudot and AMTOR systems the 5- or 7-bit code needs to be translated to ASCII, which the computer can understand. This can be done by using a look-up table in memory. Here the received code number gives the location of the ASCII code version in the stored table.

Computer programs are available for this type of direct decoding of RTTY and AMTOR signals for most popular computers, such as the Sinclair Spectrum, Commodore 64, Amiga, Atari and IBM PC compatible machines. One computer which is likely to be popular for this type of operation is the Atari Falcon, which uses a 32-bit 68030 microprocessor and also contains a 51001 digital signal processor (DSP). Audio input is converted into digital form within the computer and may be fed to the DSP chip, which can then perform the filtering and demodulation before passing digital data to the main processor. The serial output codes can also be converted into tones by the DSP chip and are then converted to analogue signals and output on an audio output line. For those with an IBM compatible PC it is likely that similar DSP circuits may become available as plug-in boards.

Where to find amateur RTTY

The frequencies generally used for amateur RTTY on the HF, VHF and UHF bands are:

Band	Frequency
3.5 MHz	3580–3620 kHz
7 MHz	7035–7045 kHz
10 MHz	10 140–10 150 kHz
14 MHz	14 080–14 100 kHz RTTY
	14 075–14 080 kHz AMTOR
18 MHz	18 100–18 110 kHz
21 MHz	21 080–21 120 kHz
24 MHz	24 920–24 930 kHz
28 MHz	28 050–28 150 kHz
144 MHz	144.600 MHz (FSK)
	145.300 MHz (AFSK)
430 MHz	432.600 MHz (FSK)
	433.300 MHz (AFSK)

For the frequencies of commercial RTTY stations see the section on utility stations in Chapter 8.

RTTY bulletin boards

One application of RTTY-type transmissions is the provision of automatic stations which act as mailboxes or bulletin board systems (BBS). This type of station can accept RTTY messages from other stations and store them in its computer memory. The messages may be addressed generally to anyone using the board or may be directed to a particular station. When you access a bulletin board of this type it will usually tell you if there are any messages for your station.

To log in to the mailbox or BBS the amateur station usually has to send its callsign and CR/LF when requested by the BBS station. On receipt of this signal the BBS responds by acknowledging the station call and giving instructions on the commands for operating the board. Most boards will provide a list of messages currently on file and may also give a log of stations that have accessed the board during the previous 24 hours. The amateur station may send a message to be stored and forwarded by the board or for general display to any amateur accessing the board. Some boards provide other facilities such as conversion of locator codes, details of the station equipment and perhaps simple computer programs which can be run by sending appropriate commands.

Most of the original RTTY bulletin boards have now changed to packet radio but a few can still be found using AMTOR and RTTY:

G3PLX	3587 kHz	AMTOR BBS
W2RJK	14 092 kHz	RTTY mailbox
W4VYU	14 092 kHz	RTTY mailbox

The PacTor system

Whilst AMTOR works very well, particularly under difficult conditions, it is a rather inefficient method of transmitting data in comparison with HF packet radio. In the ARQ mode the character rate is about 6 per second. On the other hand, HF packet radio can give a rate of about 20 characters per second but becomes very inefficient under fading or noisy conditions because of the number of corrupted packets and resultant signal repetitions. Packet radio also becomes very slow when several stations are operating on the same frequency.

In an attempt to overcome the limitations of AMTOR and HF packet, some amateurs in Germany carried out experiments using a new transmission system based on AMTOR but using longer bursts of data. This new mode of operation is called PacTor and can provide better and faster transmission on the HF bands than conventional AMTOR, especially under unfavourable reception conditions. The system has also been made flexible so that it can be used as an alternative to packet on the HF bands.

PacTor can operate at two different data rates according to the conditions prevailing on the radio link. The normal speed is 100 baud but if conditions are favourable the system switches automatically to operate at 200 baud.

In the normal 100-baud mode each packet is 112 bits or 14 bytes in length and takes 1.12 seconds. The repetition rate is one packet every 1.44 seconds, giving a gap between packets of 320 ms, during which the receiving station sends its acknowledgement signal in the same fashion as for AMTOR ARQ mode A. For the 200-baud transmission rate each packet contains 224 bits or 28 bytes. When the signal is heard it sounds a bit like AMTOR ARQ but with a slower repetition rate.

Each packet starts with a header byte which provides synchronization and can also be used to determine the phase of the FSK signal. This header byte consists of alternate 1 and 0 bits. The data field consists of 10 bytes (80 bits) and may contain ASCII character codes or be treated as simply a stream of binary bits. The data packet is completed by sending a status byte and two CRC bytes to provide error checking. This is shown in Figure 5.11. In the 200-baud mode the data field contains 192 bits or 24 bytes.

At the start of a contact the transmitting station sends a special packet which contains the callsign of the station being called (Figure 5.12). This callsign takes up eight bytes, which are sent at 100 baud, and is followed by the same data transmitted at 200 baud. The receiving station then sends an acknowledgement signal if it has received the 100-baud data correctly. If the 200-baud data have also been correctly received, a different acknowledgement signal is sent and the mode of transmission then switches to 200 baud. Once contact has been established, the transmission sequence follows the same basic scheme as for AMTOR, with the receiving station requesting repeats when it receives a corrupted packet. This follows the normal

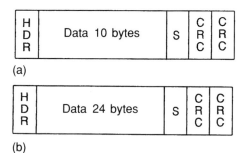

(a)

(b)

Figure 5.11 The basic frame format for PacTOR signals: (a) 100-baud transmission rate; (b) 200-baud transmission rate

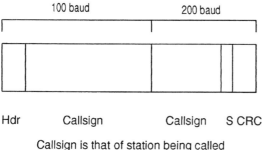

Callsign is that of station being called

Figure 5.12 The PacTOR frame format used for making an initial call to another station

AMTOR procedure of sending alternate CS1 and CS2 codes when the data are correct but repeating the last code if an error is detected. To change the direction of transmission the receiving station sends a CS3 code as an acknowledgement of the received packet.

For even more efficient communication a mode known as memory ARQ may be used. In this scheme corrupted versions of a packet are stored in memory and compared to eliminate the bit errors, so that from a few copies of the packet with errors an error-free packet may be derived, even though none of the repeated packets was error free. This involves some digital signal processing and can only be provided in hardware. By using the memory ARQ mode the data can be successfully transmitted even in adverse conditions where virtually every packet is corrupted.

Another development which can be used in PacTor is the use of a Huffman code character set, where the character length may vary from two to seven bits. Thus characters such as e, n and i might consist of only a 3-bit code, whilst less frequent characters, such as q and z, would be seven bits long. This is similar to the principle used in Morse code. This type of coding provides some degree of data compression so that a typical character length becomes around 4.5 to 5 bits, instead of the normal seven bits used for ASCII. The use of such coding schemes does require accurate decoding of the received bit stream.

The official PacTor controller unit can be obtained as a kit of parts or completely built from the German designers or from agents in other countries. It is possible to produce software for use in some home computers or packet terminal node controllers (TNCs) which will provide a basic version of the PacTor system. At present only the official hardware is likely to be able to support the more advanced feature such as memory ARQ. Variations of the system also seem to be under development and some transmissions have been heard which seem to use a shorter data block and faster repetition cycle of about 1 second.

6

Packet radio

The latest form of digital communication now being used by amateurs is packet radio. The system effectively sets up a digital communications network between two or more stations. Each station may send packets or frames of data over the link and these may be addressed to a specific station or broadcast generally on the net. The system used is called AX25 and is a variation of the X25 data link protocol which is used for digital communications on telephone networks.

Packet radio uses synchronous transmission where character codes are sent one after another with no start or stop bits. A special flag character is sent at the start of the data block to provide a timing reference from which the timing of all other data within the block may be derived. The 8-bit ASCII character code is used. Each frame consists of a series of 'fields' which may contain one or more 8-bit data codes.

The system is designed to operate as a network so that stations may simply receive frames and then forward them on to another station in the chain until eventually the frame reaches the destination station. The linking stations may be digital repeaters, or forwarding may be done via another amateur station which acts as a network node. In a complete message sent via a network of stations some frames may travel along different routes between the source and destination stations. In the header section of the frame a serial number code indicates the order in which the received frames should be reassembled to reproduce the complete message.

Packet frame format

The basic frame format for amateur AX25 signals consists of a flag field, an address field, a control field, the main data block, an error check field and a final flag to indicate the end of the frame. This is shown in Figure 6.1.

Start flag	Address field	Control field	Information (text) field	F C S	End flag
1 byte	14 – 70 bytes	1 byte	Up to 256 bytes	2 bytes	1 byte

Figure 6.1 The basic format for a frame of information in packet radio transmissions

Flag fields

The starting and ending flag fields have the unique data pattern 01111110. The start flag is used for synchronization and defines the point from which the timing of all following data words is derived. To avoid this pattern occurring in other fields the transmitter automatically inserts a 0 bit after any string of five 1 bits during the other fields of the packet frame. This is known as 'bit stuffing', and these extra 0 bits are removed automatically by the packet decoder at the receiver end.

Address field

This field contains 14–70 bytes of address information arranged in groups of seven bytes each. In each group the first six bytes hold the station callsign in upper-case letters and figures, and the seventh byte is a secondary station identifier (SSI) used to differentiate between stations, such as digipeaters and bulletin boards, which may use the same callsign. The first 7-byte group identifies the station to which the message is being sent whilst the second

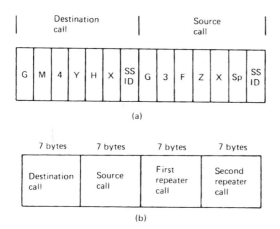

Figure 6.2 The packet radio address field formats: (a) for simple two-station contact; (b) for connection via digipeaters

ASCII character code

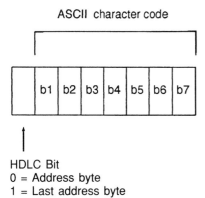

HDLC Bit
0 = Address byte
1 = Last address byte

Figure 6.3 The modified bit layout used for the address field bytes in packet radio

group identifies the source station which is sending the message. This is shown in Figure 6.2.

In the address field the ASCII code occupies the last seven bits of the byte and the first bit is normally set at 0, except in the last byte of the address field, where it is set at 1. The ASCII code is sent with the least significant bit first. This arrangement is shown in Figure 6.3.

If the message is to be routed via digipeaters, up to eight further addresses of seven bytes each may be added to identify the stations through which the message is to be passed. These digipeater calls are sent in sequence, starting with the digipeater nearest the sending station. Most packet transmissions today use only two groups in the address field.

Control field

The control field consists of one byte which indicates the type of frame being sent. The three types of frame are information (I), supervisory (S) and unnumbered (U) (Figure 6.4).

For I frames the control field contains two 3-bit sequence numbers which are used to keep track of which frames have been transmitted and successfully received.

Supervisory frames are used to acknowledge receipt of I frames or to request retransmission. Bits 2 and 3 of the control field indicate the action required and the sequence number indicates the last frame number that was successfully received.

Unnumbered frames are used for various control functions. Bits 2, 3, 5, 6 and 7 of the control field in these frames are used to indicate the type of action required. Unnumbered information (UI) frames may be used for net or round table operation but errors in received data do not cause retransmission of frames.

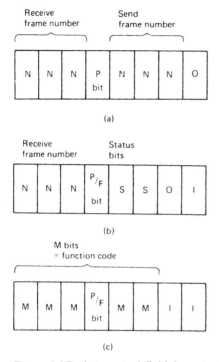

Figure 6.4 Packet control field formats: (a) information (I) frames; (b) supervisory (S) frames; (c) unnumbered (U) frames

Information field

The information field contains the message that is being sent and may be up to 256 bytes in length. This field may also contain control information in the first few bytes which indicate the type of network protocol that is in use. Text is transmitted as ASCII codes which occupy the first seven bits of each byte, with the last bit being used as a parity bit. For data transmission, such as a computer program, the field is treated as a stream of data bytes rather than character codes and various special data formats may be used according to the data application. The type of data protocol being used is indicated by the first few bytes of the field.

Frame check sequence

The frame check sequence (FCS) is a 16-bit word which provides an error check on the data sent within the frame. The transmitted FCS word is calculated from the data sent in the transmitted frame. At the receiving end a similar FCS word is calculated from the received data and if this matches the received FCS then the data are correct and are accepted. The FCS word occupies two bytes and is followed by a flag field which completes the packet frame.

AX25 protocol specification

For more details of the full AX25 protocol used for packet radio, readers should consult *The AX25 Specification* published by ARRL and available in the UK from the RSGB.

Transmission technique

At each station the encoding and decoding of frames and their assembly or disassembly to provide messages is controlled by a small computer called a terminal node controller (TNC). In most cases this is a dedicated microprocessor-based device but it would also be possible to use a personal computer to perform this function, provided it were fitted with a suitable serial interface capable of handling the HDLC-type data link protocol. The computer would also need software to decode the packet address field and to reassemble the received packets.

Transmission uses FSK or AFSK with a shift of 200 Hz on the HF bands and 1000 Hz on the VHF or UHF bands. Data are usually converted into the non-return to zero space (NRZ-S) format before transmission. In this scheme a 0 in the data bit pattern causes a change in output frequency, whilst a 1 bit causes no change (Figure 6.5). For packet transmission via satellite, phase shift keying (PSK) is often used.

TNC types

Most amateurs operating packet tend to use commercially built TNCs together with a home computer to provide their packet facility. There are a number of popular TNCs available and here we shall look at some of their features.

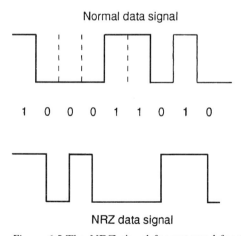

Normal data signal

1   0   0   0   1   1   0   1   0

NRZ data signal

Figure 6.5 The NRZ signal format used for transmission of packet radio data

TINY2

This widely used TNC is the TINY2 Mk II, which is made by PacComm and is perhaps the least expensive commercial TNC. This unit is based around a ZX80 processor and is designed for VHF and UHF packet only. It runs from 12 V DC and can be used as a node or digipeater.

PK88

Another economical TNC is the AEA PK88, which is also based on the Zilog Z80 and requires a 12-V DC supply. This unit will operate on HF, VHF and UHF packet and includes a personal mailbox with 18 kilobytes of memory. This unit can also be used to implement level 3 and 4 networking using NETROM.

PK232

Another popular TNC from AEA is the PK232MBX, which provides multi-mode operation. This unit can handle HF/VHF packet, Morse, RTTY, AMTOR, WEFAX and NAVTEX operation and can also be used to receive and print HF FAX weather charts. This unit uses a filter/discriminator circuit in the modern section which gives better performance than a phase-locked loop demodulator in noisy conditions. The unit also includes automatic signal recognition which can detect the mode and speed of received RTTY/AMTOR signals and select the appropriate mode. The internal software is compatible with TCP/IP and includes an 18-kilobyte mailbox.

TNC320

This unit from PacComm provides HF and VHF operation with two separate modems. The software of this unit supports TCP/IP and provides a 12-kilobyte personal mailbox. Based on a Zilog Z80 processor with a TCM3105 modem IC, this unit runs from a 12-V DC supply.

KAM

This unit produced by Kantronics is a multimode controller similar to the PK232. It can operate using CW, RTTY, AMTOR, ASCII and HF/VHF packet. The software includes KA-NODE and BBS operation.

DSP1232 and DSP2232

These two units produced by AEA are perhaps the ultimate in amateur data mode controllers. They use a Motorola 56001 digital signal processor for filtering and modem functions and a Zilog 64180 processor for digital operations. The unit can handle CW, RTTY, AMTOR, ASCII, HF/VHF packet, satellite PSK packet, WEFAX, FAX and SSTV. The 2232 provides two simultaneously operating I/O ports whilst on the 1232 the ports are switchable.

Using a home computer

If you have a packet TNC the home computer is simply used as a terminal to organize the display of packet messages and to provide keyboard input.

Many home computers can be programmed to perform the function of a TNC as well as displaying text and providing keyboard input. In this case a simple external modem is used to convert the incoming audio into a digital data signal which is then fed into one of the input ports of the computer. Similarly, a serial data signal is output by the computer to the modem, where it is converted to audio tones which are fed to the audio input on the transmitter. The computer software used will determine what TNC commands and facilities are available.

A widely used modem chip for VHF and UHF packet is the TCM3105. This chip contains a digital filter circuit at the input to clean up the audio signal and a phase-locked loop demodulator circuit to extract the digital signal from the audio tones. The digital signal for the transmit message is fed to a two-tone generator within the chip which produces the required audio tone for the transmitter. The basic circuit for the modem is shown in Figure 6.6.

A few home computers, such as the Atari Falcon, have the capability of performing the functions of modem and TNC with suitable software. For the IBM compatible PCs it is possible to buy, or build from a kit, a plug-in card which will perform the functions of TNC and modem.

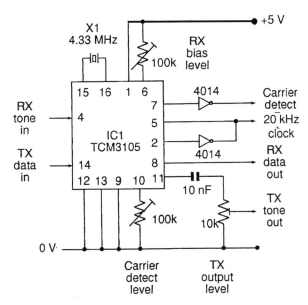

Figure 6.6 Modem circuit for VHF packet radio using the TCM3105 integrated circuit

Getting started with packet

Having obtained a suitable TNC and connected it to the transceiver and computer, or terminal, the first stage will be to set up the TNC. The TNC will usually start initially with a set of default settings which will often be satisfactory for initial experiments with packet. Some suppliers will program in your callsign before sending the TNC, but if this has not been done the callsign will have to be programmed using the command MY followed by the station callsign. The CW ident must also be set up using the command CWID followed by the station callsign and the timing by using the command CW *nn* where *nn* is a number giving a delay time. This should be set for either 15 or 30 min. Most TNCs will not transmit until the callsign and CW ID settings have been completed.

Most TNCs will automatically have the monitor mode switched on after power up; so tune in to a local packet channel and the screen should start to display packets being sent by other stations. If the monitor mode is not on, use the command

mon on <return>

Once the monitor mode is running, it may be helpful to use the command

mcom on <return>

since this will cause command and response packets to be displayed as well as those which contain text. Whilst the monitor mode is selected, the audio level from the receiver can be adjusted for reliable operation.

Having got the TNC reading received packets, the command

mheard <return>

command can be used. This will cause the TNC to list the callsigns of the last 10 or 20 stations that it has heard. By using this command it is possible to find out which stations are within range. In many cases the TNC software will indicate those stations which are acting as nodes or BBS stations. If a node is relaying signals from a BBS it may also be tagged as a BBS on the MHeard list. BBS stations usually send an identification frame from time to time, so that callsigns of BBS stations can be determined from these identification frames. After monitoring the local channels for a while, the next stage is to actually make a connection to one of the stations that can be heard.

After switch-on, the TNC goes into COMMAND mode, where it will treat all input from the computer keyboard as commands. When a connection to another station is made, the TNC switches to CONVERSE mode, where it will send input from the keyboard as packets to the other station. Each time a carriage return is keyed in, the previously entered text is sent as a packet. Alternatively the TNC will send a packet automatically after it has received a specified number of characters. To return to the COMMAND mode, a control code is input on the keyboard by pressing Control and C simultaneously. This is the default code, but most TNCs will allow you to change it to a different code if Control C causes problems with the computer software.

To connect to a station use the command

c G3xxx<return>

assuming that G3xxx is the callsign of the station you are connecting to. The TNC will attempt to make the connection, and if successful it will print the message

Connected to g3xxx

and then switch to CONVERSE mode so that any message you type in will be sent to g3xxx. If the connection attempt fails, the TNC will make a further nine attempts and then print an error message. To disconnect you will need to return to the COMMAND mode and then type in the command

d <return>

HF packet frequencies
On the HF bands a number of frequencies are used for packet radio as follows:

Band (m)	*Frequency* (kHz)
80	3590–3600, BBS on 3605
40	7035–7045
30	10 140–10 150, BBS on 10.1453, 10.173
20	14 060–14 099, 14 101–14 110, BBS on 14 102, 14 103, 14 108
17	18 100–18 110, BBS on 18 106, 18 108
15	21 100–21 120, BBS on 21 096, 21 098, 21 103, 21 107
12	24 920–24 930, BBS on 24 926, 24 928 (USA)
10	28 120–28 150, BBS 28 123, 28 127 (Europe), 28 102, 28 104 (USA)

For HF band operation stations generally use an SSID of –3. For nodes and bulletin boards a number in the range 30–38 may be used to indicate the band in use as follows:

30–1.8 MHz; 31–3.5 MHz; 32–7 MHz; 33–10 MHz; 34–14 MHz; 35–18 MHz; 36–21 MHz; 37–24 MHz; 38–28 MHz

All HF packet stations use 300 baud with a 200-Hz frequency shift.

VHF packet frequencies in the UK
On the VHF and UHF bands several frequencies are used for packet radio. In the following list the number in brackets is the channel identifier which may be added to an alias identifier.

Six-metre band (SSID –6)
50.61 MHz (67), 50.63 MHz (60), 50.65 MHz (61)
50.67 MHz (62), 50.69 MHz (63), 50.71 MHz (64)
50.73 MHz (65), 50.75 MHz (66)
50.69 MHz is used for a 9600-baud packet.

BBS are usually on 50.65 and 50.67 MHz.

Four-metre band (SSID –4)
70.3125 MHz (40), 70.3250 MHz (41), 70.4875 MHz (42)

Two-metre band (SSID –2)
144.6125 MHz (2A), 144.625 MHz (20)
144.650 MHz (21), 144.675 MHz (22)

BBS are usually on 144.650 MHz.

UHF packet frequencies in the UK

70-cm band (SSID –7)
430.625 MHz (76), 430.675 MHz (77), 430.700 MHz (78)
430.725 MHz (79), 430.775 MHz (87), 432.625 MHz (70)
432.650 MHz (71), 432.675 MHz (72), 433.625 MHz (73)
433.650 MHz (74), 433.675 MHz (75), 438.050 MHz (7A)
439.775 MHz (80), 439.825 MHz (81), 439.850 MHz (82)
439.874 MHz (83), 439.900 MHz (86), 439.925 MHz (84)
439.975 MHz (85)

23-cm band (SSID –1)
1240.150 MHz (10), 1240.300 MHz (11)
1240.450 MHz (12), 1240.600 MHz (13)
1240.750 MHz (14), 1270.000 MHz (1H)
1278.000 MHz (1J), 1286.000 MHz (1A)
1299.000 MHz (15), 1299.100 MHz (1B)
1299.425 MHz (16), 1299.575 MHz (17)
1299.650 MHz (1D), 1299.675 MHz (1E)
1299.725 MHz (18), 1299.959 MHz (1F)

12-cm band
2355.00–2365.00 MHz
2392.00–2400.00 MHz

Frequencies used are 2310.1, 2310.3, 2355.1, 2355.3 and 2364.1 MHz

Digipeaters
One problem which can occur particularly with packet radio on VHF and UHF is that only stations within local radio range can be connected. This can be overcome by using a digital repeater (digipeater), which receives the signal from the originating station and retransmits it, usually on the same frequency. To send the signal further afield it is possible to link through up to eight digipeaters.

In the early days of packet radio, digipeaters used GB3 callsigns and needed a special licence. Today most packet TNCs have a digipeat mode which allows the TNC to act as a digipeater. Some stations indicate that they have digipeat (DIGI) on when they send their identification messages.

To link through a digipeater the connect command specifies the destination callsign and the digipeater(s). The command takes the form:

c g1abc via g3xyz, g0uvw <return>

where g1abc is the station you wish to communicate with, g3xyz is the first digipeater and g0uvw is the second digipeater. Up to eight digipeaters can be included in the list but these must be in sequence, starting with the digipeater nearest the sending station.

Today most stations would use network nodes rather than digipeaters to achieve connections to more distant stations, since this provides a more reliable communications link.

Network nodes
The problem with using digipeaters to reach distant stations is that the digipeater simply relays whatever it receives without checking for errors. The error checks are carried out at the destination station, so if an error occurs anywhere along the chain of digipeaters the packet will be rejected and will have to be sent again.

A better system of sending messages to distant stations is by using network nodes. Unlike a digipeater, the node will carry out error checks before passing on its messages and provides a more reliable link. Many amateur stations have added network node software to their systems and leave their stations on air unattended to act as a link in the packet network.

To use a network node the first step is to connect to the node station using a normal connect command. Node stations often use an alias as a callsign but a connection may be made by using either the alias or the actual callsign. Once the connection is made, the node usually sends back a welcome message and perhaps a list of available commands. At this point your terminal is effectively sending commands to the node software.

Once a connection is established with the node it is possible to connect to any other node or station that is within range of that node. This process can

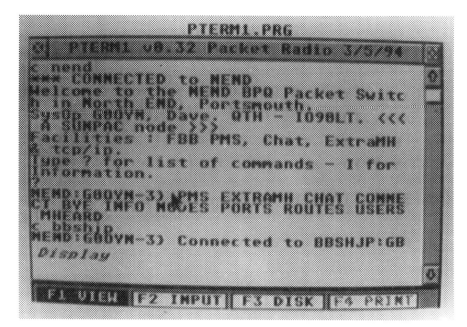

Figure 6.7 Example of a typical connect message from a packet radio network node

be repeated until the desired station is reached. Unlike a digipeater link, where only eight digipeaters can be used, a link via nodes can use any number of nodes.

One useful command is NODES, which can cause the node station to send a list of other nodes that it can contact. Another command on some nodes is ROUTES, which will send a list of routes to more distant nodes showing the various nodes that you will need to pass through to reach the desired station. To progress from the node to the next point in the network you simply enter a connect command to the next node you wish to link into. The process is then repeated from the new node until you reach the station that you are aiming for. Unlike digipeaters, the node stations may make the connection to the next node in the chain on a different frequency or a different band, so that it is possible to link around the world by using this technique.

There are various types of software used for node operation. The simpler and perhaps more common type is the KA node. This type of node requires that the originating station should sort out the links through the network.

Commands available on these nodes are:

Command	Function
Abort	Abort last command
Bye	Disconnect
Nodes	List of nodes that are in range
Routes	List of routes to distant stations
JHeard	Lists stations recently heard
Help	Help on using the node
Info	Details of node station
Connect	Command to connect to another node

Help regarding commands can be obtained by sending ? followed by the command word; the node will send back a message explaining the command. In most cases only the first letter of the command need be sent.

The second type of node is the NETROM type. The difference in this type of node is that the routing to a more distant station is done automatically by the NETROM software. This has the advantage that the software can choose the best route, taking into account traffic and propagation conditions.

The command set for a NETROM node is similar to that for a KA node but with some different commands:

Command	Function
Users	Lists users currently on the node
Nodes	Lists nodes and routes
Ident	Identifies the node
CQ	Sends a CQ message
PARAMS	Sends details of the node station

Some nodes have a command PORTS which lists the various ports available on the node and their frequencies. By specifying a different port when you send a connect command to the node it is possible to connect to a node or BBS which is operating on a different band.

Personal mailboxes

Most TNCs have a personal mailbox facility which will allow other stations to leave messages for you when you are not actually present at the station. For this to work you will need to leave your TNC and station operating unattended whilst you are away. The TNC will automatically send an identification message at regular intervals with a message giving your mailbox callsign or alias. The call or alias will usually have the SSID set as –2.

If a station wishes to leave a message, he or she connects to your mailbox and follows the instructions for leaving a message. On your return you can check if any new messages have arrived and call them up to the display for reading.

Bulletin board systems (BBS)

A BBS is a sort of public mailbox where stations can leave messages for one another but it will usually include many other information files which can be accessed. Apart from text messages there may be computer program files or picture files. To operate a BBS a special licence is required and the station callsign in the UK will be in the GB7 series.

Various types of software are available for use in controlling bulletin boards, including WORLI, AA4RE, F6FBB, MSYS, NNA, G4WSD, AP-MAIL and TCP/NOS. These have slight differences in their command sets, but, by using a HELP command, the BBS will usually provide a listing of the commands required to operate it.

To connect to a BBS turn off the monitor mode and send the command

c gb7xxx<return>

using the callsign of the BBS. The TNC will then try to connect to the BBS station, and if it is successful the message CONNECTED TO GB7xxx should appear. This will usually be followed by a greetings message from the BBS and brief instructions on how to use the BBS. The first time you connect to a BBS it will often ask for your name and your locator code, which is then stored in the BBS computer so that it will recognize you when you call in

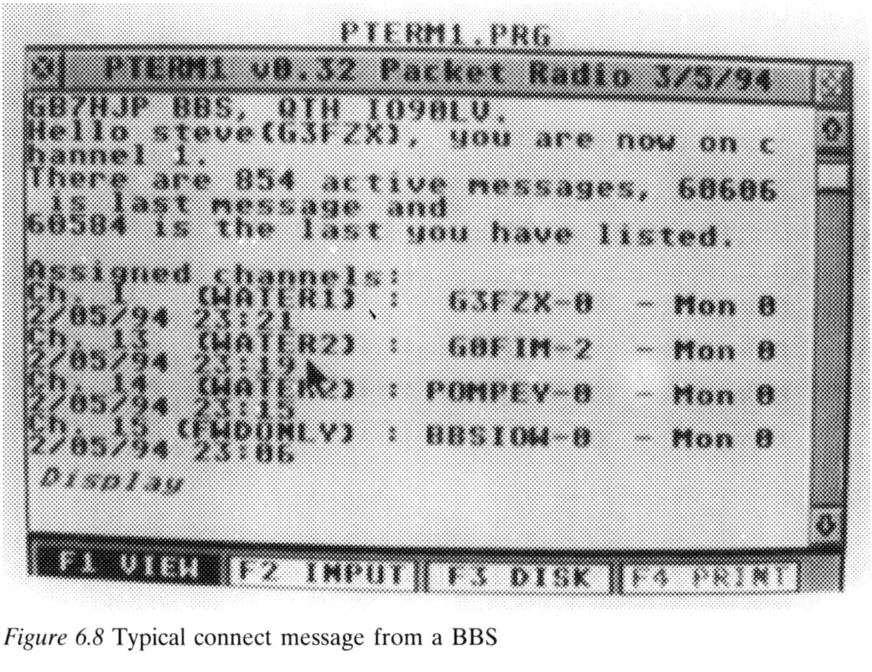

Figure 6.8 Typical connect message from a BBS

again. If there is no BBS in direct range you may need to make the connection to the BBS via a node or digipeater.

The bulletin board will usually tell you if there are any messages addressed to you. There are often thousands of messages stored in a bulletin board at any given time. Many of these are bulletins which are addressed to *all* users and may give information for satellite orbit parameters to cookery recipes. Personal messages for a specific station can usually be read only by that station.

The list of bulletin messages available can usually be obtained by using the command

lb<return>

Each message in the list will have a message number, and to read a message use the command

r nnnn<return>

where nnnn is the message number.

Messages for other amateurs around the world can be passed via the local BBS, which then sends them on through other BBS stations until they eventually arrive at the local BBS of the station to whom the message is addressed. You will need to specify the target BBS and give your own home BBS call so that a reply can be sent back to you. Sometimes you may need to specify the route through the network by listing the callsigns of the various BBS stations along the path.

F6FBB BBS command set

As an example of the commands used on a BBS, the following list shows the command set for the F6FBB-type bulletin board software which is widely used.

Command	Function
A	Abort listing
B	BYE (log off and disconnect)
C	Enter conference mode
D	Access to FBB DOS
F	Access to file server
G	Gateway to other frequencies
H	Help
I	Details of the BBS system
J	List recently heard stations
JK	List last 20 heard stations
K nnn	Kill message number nnn
L	List new messages
LB	List bulletins
LL 10	List last 10 messages
LM	List messages to you

Command	Function
LN	List unread messages to you
LS text	List messages containing text
L@	List messages addressed to the BBS
L> G3XXX	List messages to callsign
L< G3XXX	List messages from callsign
N name	Change your name on BBS file
O	Selects options
R nnn	Read message number nnn
RM	Read messages addressed to you
RN	Read new messages addressed to you
S G3XXX	Send message to G3XXX
SP G3XXX	Send personal message to G3XXX
SB	Send bulletin to all stations
T	Talk to system operator
V	List version of BBS software
X	Select expert mode
Y	Select YAPP binary file transfer

Note that you can only use the K command to kill messages that are addressed to your station. The callsign G3XXX is given as an example but you would use the actual callsign of the station you wish to send a message to or list a message from. The message number nnn and message title are given in the message listing that you receive after an L command. New messages to you are ones that you have not read yet.

YAPP is a binary file transfer protocol which can be used for transferring data files and computer programs. You will require suitable software in your TNC or computer program before you can use this form of data transfer.

Dx clusters in the UK

Another form of BBS is the Dx cluster station, which can provide up-to-date information on Dx stations, satellites and so on. Apart from the normal BBS activity, the Dx cluster provides a conference-type mode where many stations are all connected simultaneously and can pass information to one another regarding Dx activity in real time.

Dx clusters in the UK at the time of writing are:

GB7BPQ (NMCLUS)	Mapperley, Nottinghamshire 61, 41, 20, 72, 18
GB7DXC (DXCHEL)	Cheltenham 41, 21
GB7DXH (DXH)	Hemel Hempstead 62, 65, 41, 22, 75
GB7DXI (CHILDX)	Wokingham, Berkshire 41, 22, 73, 18
GB7DXM (DXMCLS)	Martlesham, Suffolk 22, 72
GB7DXS (DXSUSX)	Haywards Heath, Sussex 41, 1D
GB7LDX	Wallasey
GB7MDX (DXCHES	Stockport, Cheshire 41, 22, 72
GB7PDX (PDX)	Plymouth 42, 22, 72

GB7SMC (SMCDX)   Chandlers Ford, Hampshire 61, 41, 22, 72
GB7TLH   East Dereham, Norfolk 2A, 22, 72
GB7WDX (BRIDX)   Bridgwater, Somerset
GB7YDX (DXYORK)   Wetherby, Yorkshire 41, 22, 81, 7A

TCP/IP network

Many packet transmissions now use the TCP/IP protocol to address and send messages. This protocol was originally developed for use on US military data networks and has been adapted for amateur use. Each station has a unique TCP/IP address code which specifies the country and region. When a message is sent to another station using this system, the route to the other station is sorted out automatically. The amateur radio section of the global TCP/IP network is called AMPRNET.

The TCP/IP address is a 32-bit binary number which is divided into four bytes. For amateur radio the first byte has the value 44. The second byte identifies a subdomain, and for the UK this has the value 131. The third byte is used as a regional address and in the UK this is the number of one of the 20 RSGB regions, with the Isle of Man and the Channel Islands being designated as regions 21 and 22 respectively. The final byte identifies the individual station. Thus a TCP/IP station in the UK might be allocated an IP address of 44.131.5.20, indicating that it is in RSGB region 5. Stations wishing to use TCP/IP must first be allocated an address and this is handled by applying to G6PWY at the GB7PWY BBS.

To address a UK TCP/IP station the address used is written in the form

callsign.region.uk.amp

for example:

g3xyz.r6.uk.amp

The station with the TCP/IP address may act as a host for other local stations to access the TCP/IP network. In this case they would be addressed as

g3abc@g3xyz.r6.uk.amp

To operate directly on the TCP/IP network the net operating system (NOS) packet software is required.

Gateways

A mode of operation which may be used in packet radio is the gateway system. This is a form of node which provides an interconnection between VHF or UHF packet and HF packet, AMTOR, or amateur satellite links to another remote packet network.

Thus it is possible for a class B VHF packet station to send messages via an AMTOR HF gateway to other packet stations anywhere in the world. Similarly, by using a satellite gateway the same result can be achieved.

Some satellites are equipped with store and forward facilities so that messages for stations not currently in range of the satellite can be sent when the satellite comes into range of the destination station. The possibilities of packet radio as a communications medium are extensive and many avenues are being explored for possible future use.

RS232 connections

The normal link used between a TNC and the personal computer uses a serial data link and may have either a 25-way D connector or a 9-way D connector at the computer end. The signals at the computer end are as follows:

9-pin D connector
1   Data carrier detect (DCD)
2   RX data (input)
3   TX data (output)
4   Data terminal ready (DTR) (output)
5   Signal ground
6   Data set ready (DSR) (input)
7   Request to send RTS (output)
8   Clear to send CTS (input)
9   Ring indicator RI (input)

Shell is frame ground

25-pin D connector
1   Frame ground
2   TX data TXD (output)
3   RX data RXD (input)
4   Request to send RTS (output)
5   Clear to send CTS (input)
6   Data set ready DSR (input)
7   Signal ground
8   Data carrier detect (DCD) (input)
9   –
10 –
11 –
12 –
13 –
14 –
15 –

16   –
17   –
18   –
19   –
20   Data terminal ready DTR (output)
21   –
22   Ring indicator RI (input)
23   –
24   –
25   –

RTS is sent by the computer to indicate that it is ready to accept data on the RXD line. CTS is used by the TNC to tell the computer that it is ready to receive data from the TXD line. These signals may be used to control data flow between the computer and the TNC. Pins 9 and 10 on the 25-way connector may be used to carry +12 V and –12 V signals for test purposes with some TNCs.

The data and control signals on an RS232 system are –12 V for mark and +12 V for space, although many systems will operate on signals as low as +3 V to –3 V.

7

Picture transmission

Although most communications systems use voice or telegraphy for the transmission of messages, it is also possible to transmit pictures via radio. The most familiar of these techniques is television (TV), which provides high-definition moving pictures. TV requires the use of wide-bandwidth channels and is usually restricted to the UHF and microwave bands. Pictures can be transmitted using narrow-band radio transmissions on the short-wave radio bands by using special techniques, such as FAX and slow scan television.

Amateur fast scan television (FSTV)

Amateur transmissions of conventional high-definition TV both in monochrome and in colour may be found in the 432-MHz and 1200-MHz bands and also in the higher bands such as 2300 MHz. Transmissions usually follow the same standard as that in use in the country of operation.

In the UK the 625-line 50 fields per second standard would be used with PAL encoding for colour transmissions. North American amateurs use the 525-line 60 field per second system with NTSC coding for colour.

On 432 MHz the transmission method uses conventional AM but for the higher bands FM is often used for the video signals. Sound is generally transmitted on a separate channel, often on 144 MHz, but some transmissions may use intercarrier sound.

FSTV on the 432-MHz band is generally confined to the upper end of the band from 434.00 to 440.00 MHz, which is also shared with the amateur satellite service. Amateur television (ATV) operation is normally arranged to avoid causing interference to other users of this part of the 432-MHz band. In most cases only the vision signal is transmitted in the 432-MHz band, but a sound link for talkback is usually established on the 144-MHz band which allows the receiving station to report on the quality of the received pictures and the transmitting station to comment on what is being transmitted on the vision channel. The generally used frequency for ATV talkback is 144.750 MHz, and this channel is also used to establish communications before a television transmission is attempted.

When the 1300-MHz band is used for FSTV transmissions, either AM (A3F) or FM (F3F) may be used for the vision signal. The voice signal may be transmitted using FM (F3E) at a frequency 6 MHz above the vision carrier. For AM video transmission this may be a 6-MHz subcarrier. Colour transmissions are made using the PAL system with chrominance information transmitted on a subcarrier of 4.33 MHz. Colour cameras are now readily available for use with home video recorders, and a suitable video signal for the transmitter is generally available from a video recorder. Previously recorded material may be sent by using the playback signal from a video recorder to drive the vision transmitter. Amateur stations are not permitted to broadcast entertainment programmes, so any pre-recorded material used must consist of amateur videos of family or radio club activities and local events.

FM fast scan televison (FMTV)

Although normal FSTV signals use AM for the video component, an alternative approach is to use FM for the video signal. This takes up a wider bandwidth but provides some advantages and is now used by amateurs for transmissions on the 1200- and 2300-MHz bands. This mode of transmission is similar to that used for commercial satellite TV transmissions but with a narrower deviation in order to conserve bandwidth.

Typically a peak-to-peak FM deviation of 8 MHz would be used and the receiver IF (intermediate frequency) would be fitted with a filter giving 16-MHz bandwidth. It is possible to modify some satellite TV receiver IF strips for use in amateur FMTV receivers. For FMTV in the 10-GHz band, a peak-to-peak deviation of 16 MHz is often used and the receiver bandwidth is increased to 27 MHz. Pre-emphasis, where the higher video frequencies are boosted before being fed to the modulator, is used to improve the effective signal-to-noise performance. De-emphasis is applied in the receiver to restore the correct video signal.

Transmitters for 23-cm FMTV often use a basic transmitter operating in the 432-MHz band to drive a varactor tripler to produce the 1200-MHz signal. High-gain beam antennas can be used to produce a high effective radiated power for good signals over line-of-sight distances. For longer distance operation, TV repeaters are used.

Televison repeaters

In the UK several amateur repeaters have been set up for FSTV signals. These operate in the 1200-MHz and 2300-MHz bands using the following channels:

RMT1: input 1276.5 MHz; output 1311.5 MHz

GB3UT   Bath, Avon (AM)
GB3VI   Hastings, East Sussex (AM)

RMT2: input, 1249.0 MHz; output, 1318.5 MHz

GB3CT   Crawley, West Sussex (FM)
GB3GT   Glasgow, Scotland (FM)
GB3GV   Leicester
GB3NW   Nottingham
GB3PV   Cambridge (FM)
GB3RT   Coventry
GB3ZZ   Bristol

RMT3: input, 1248.0 MHz; output, 1308.0 MHz

GB3HV   High Wycombe

Picture quality report code

In order to report on the quality of the received television signals, the following code may be used.

P1   Barely visible
P2   Poor
P3   Fair
P4   Good
P5   Excellent

Slow scan television (SSTV)

This is a system for transmitting pictures via an HF channel using the normal AF bandwidth. This is achieved by reducing the scanning rate until the video frequency signals representing the picture information will fit into the audio bandwidth of 2.5 kHz approximately. A number of different standards are currently in use which provide monochrome or colour pictures.

The original SSTV standard used 120 scan lines with different timings for the USA and Europe because their frame rates were based on local mains frequencies.

	Europe	*USA*
Line scan period (ms)	60	66
Frame scan period (s)	7.2	8.0

The signals for SSTV are used to frequency modulate an AF subcarrier to give an audio frequency of 1200 Hz for the tip of the sync. pulses, 1500 Hz for black level and 2300 Hz for peak white. This is shown in Figure 7.1.

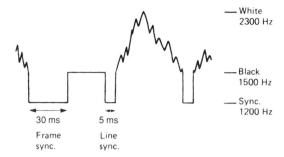

Figure 7.1 The basic waveform and timing for an 8-second black and white SSTV signal

The line synchronization pulse has a duration of 5 ms, and the field pulse takes up 30 ms of the first scan line.

Modern black and white SSTV systems

Most amateurs today use scan converter systems, such as the Robot and Wraase units, or home computers, such as the Spectrum, Commodore Amiga or Atari ST. These systems can usually handle several different modes for black and white SSTV.

The usual standards encountered are:

8 second	128 lines, 128 × 128 pixels
16 second	128 lines, 256 × 128 pixels
32 second	256 lines, 256 × 256 pixels

Robot converters can also handle 12-, 24- and 36-second modes in black and white but these modes are not often seen on the air. All black and white transmissions produce a square picture with a 1 : 1 aspect ratio.

Home computers often have only 200-line display capability in their graphics mode and are unable to show the whole picture at full resolution in the 32-second mode. This may be dealt with by storing the entire picture in memory and displaying part of it. A scrolling system is then used to allow the rest of the picture to be examined.

For good results the display should be able to handle at least 16 levels of grey, but acceptable results can be achieved with only eight grey levels. On monochrome graphics displays which can only produce black or white, the grey scales are achieved by dithering. This involves using a block of 3 × 3 or 4 × 4 pixels on the display screen to represent each pixel of the SSTV picture. By altering the pattern of dots in the block the various grey levels can be simulated, but the resultant picture has a relatively coarse, grainy structure.

SSTV in colour

Although most slow scan transmissions are of black and white images, it is possible to transmit colour pictures by using more complex transmission techniques and many stations now have facilities for handling colour pictures.

Early experiments with colour transmission involved sending three successive pictures which were taken through red, green and blue filters. At the receiving end these three pictures are stored in a picture memory which is then read out at normal FSTV speed to produce the red, green and blue video signals for a colour TV monitor, where the images combine together to produce a full colour picture on the screen. This approach is known as the field sequential colour system but it is no longer used for modern SSTV transmissions.

Line sequential colour

One of the more common colour SSTV systems is line sequential colour, in which red, green and blue (R, G and B) information is transmitted on successive scan lines of the picture. The picture is usually generated by a standard home video camera, and the composite colour video from the camera is split into red, green and blue components using similar circuits to those in a colour TV receiver.

The R, G and B images are usually stored as digital numbers in three frame stores which effectively freeze one complete frame from the video signal received from the camera. Once the frame has been captured, it is read from the frame store at SSTV speed to produce the transmitted signal. For each of the 128 lines of the SSTV picture, three successive lines are transmitted carrying the red, green and blue information respectively. If the picture is viewed on a normal black and white SSTV system, the picture is three times its usual height and the groups of three successive lines can be seen.

At the receiving end the successive lines of picture are transferred to the R, G and B planes of the frame store, and when the complete picture has been captured the stores are read out to produce R, G and B drives to a colour monitor. In most systems the output to the colour monitor is operating all of the time and the picture builds up on the screen as it is received and written into the picture memory.

In a microcomputer-based system the incoming RGB information is written one pixel at a time to the computer's screen display memory, and again the picture builds up in colour as it is received off air. When the complete picture is in memory it can readily be transferred to a floppy disk or a hard disk to provide a permanent record which can be recalled to the screen at a later date.

The usual sequential colour format is the 24-second mode, where the line scan rate is the same as for 8-second black and white, but with groups of three successive scans for each picture line so that the complete frame takes 24 seconds to transmit.

One problem with the line sequential system is that for proper reproduction of the picture the receiver must remain in step with the colour sequence of the transmission. The receiver switches between R, G and B on successive lines, starting from the first line after the frame sync. pulse. If a line sync. pulse were missed, the receiver would get out of step in its RGB sequence and the resultant colours on the screen would be incorrect.

One technique for reducing this problem is to transmit an identification pulse at the start of each red scan line. This pulse is 1 ms long at the white level and may be detected by the receiver circuit to synchronize the RGB sequence. On the G and B lines the first 1 ms of video after the sync. pulse is held at the black level. This format is sometimes referred to as the Volker–Wraase system after the German firm which introduced it in their SSTV equipment.

For higher definition pictures a 48-second mode is sometimes used. This has the same line format as the 24-second mode but has 240 scan lines to give improved vertical resolution.

A more common high-resolution line sequential mode is the 96-second mode, which uses a 120-ms line scan time and 256 lines to give a 256 by 256 pixel colour picture. Both of these systems use the red line marker pulse to ensure correct synchronization of the RGB sequence. The picture aspect ratio for line sequential colour is 1 : 1, as for black and white.

Luminance–chrominance (Robot) modes

To overcome the synchronization problems of the line sequential scheme an alternative colour transmission scheme was developed by the makers of the Robot SSTV equipment. This transmits the colour information in a single scan line and uses a multiplexed luminance and chrominance scheme. The colour picture information is split into a luminance component (Y), which is effectively a monochrome picture, and a pair of colour difference signals which convey the colour information.

There are three colour difference signals, which are known as R-Y, G-Y and B-Y. The colour difference signal (R-Y) is obtained by subtracting the luminance Y from the red video signal. Similarly, (B-Y) is obtained by subtracting Y from the blue signal. In practice only R-Y and B-Y need to be transmitted along with the Y signal, since the green signal can be derived from these three components at the receiving end. To save bandwidth the resolution of the colour difference signals is reduced to half that of the Y signal. This reduced colour resolution technique is used in normal colour TV transmissions without any serious loss of picture quality.

In the Robot system the luminance (Y) information is transmitted after the line sync. pulse. This Y signal is compressed in time to two-thirds of its normal time period. This is then followed by an identification pulse, and then a compressed version of either the R-Y or B-Y component is transmitted during the rest of the line scan. The colour difference signal is

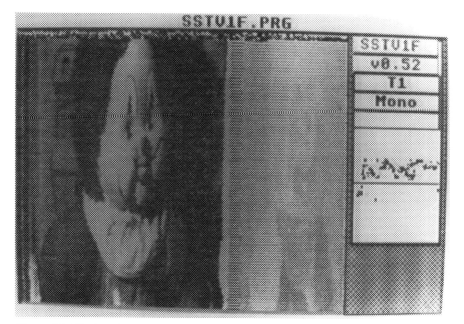

Figure 7.2 Display of a Robot colour SSTV signal showing the luminance and chrominance images

compressed to one-third of its normal period. R-Y and B-Y components are sent on alternate scan lines, and the state of the identification pulse indicates which colour difference signal is being sent. This transmission layout can be seen in Figure 7.2, which shows a Robot colour picture displayed using a monochrome mode with the same total line scan period. The luminance signal can be seen at the left of the picture and the colour difference component at the right. The identification pulses can be seen between the two parts of the picture. At the receiving end the luminance and colour difference components are expanded to their normal time periods and then combined to produce R, G and B signals which are held in the frame store and used to generate the RGB drive for a colour monitor.

The low-resolution version of the Robot colour system is generally referred to as the 36-second colour mode and has 128 scan lines. The high-resolution version is the 72-second mode with 256 scan lines. At the start of each frame there is a start signal which identifies the mode so that the converter can automatically switch to the correct mode to match the transmitted picture. Unlike the sequential colour systems, the Robot colour modes use a 4 : 3 aspect ratio, which is the same as for a normal TV picture.

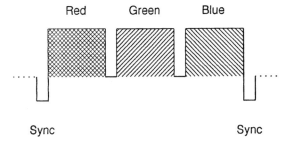

Figure 7.3 The signal format used for the Scottie mode colour SSTV pictures

Scottie mode

Another method of overcoming the synchronization problems of the line sequential scheme is the Scottie mode colour transmission system. This system was originally developed by GM3BSC. In this system the R, G and B components are sent in sequence in a single scan line as shown in Figure 7.3. The system relies on the use of crystal-controlled timing generators for the pixel scan to ensure proper synchronization of the three images in each scan line.

This scheme is much less affected by noise and interference and enables high-resolution pictures to be received even under difficult reception conditions. Even if a line synchronization pulse is missed, the effect is that one line scan may be corrupted but the system will get back into step when it detects the next line pulse.

The normal Scottie S1 mode has a line scan time of about 400 ms with a 240-line picture so that each frame takes approximately 96 seconds for transmission. There is also a Scottie S2 mode which has a frame rate of about 48 seconds. At the start of each Scottie frame there is a mode identifier signal which allows the receiver converter to select the correct mode. Because the Scottie system transmits R, G and B signals at full resolution, it gives better picture quality than the Robot colour modes.

Martin new mode

An alternative to Scottie mode which uses the same general principles is known as Martin mode. In the Martin system the colour segments on each scan line are sent in the order green, blue, red, and a slightly longer line scan time is used. Martin pictures normally have a 4 : 3 aspect ratio with 240 scan lines per frame. As with Scottie mode there is a mode identifier signal at the start of each transmitted frame. Various different Martin modes may be used but the most frequently seen is Martin M1, which gives similar results to Scottie S1.

Amiga video terminal (AVT) mode

The Amiga video terminal (AVT) mode of transmission was originally developed for use on the Commodore Amiga computer. Unlike the other SSTV modes, this one is more akin to FAX since there are no line sync. pulses. A start signal and a synchronizing sequence are sent at the start of the picture and then the scans continue synchronously. Red, green and blue line scans are transmitted in sequence in much the same way as for Scottie mode signals. As with FAX, the system relies on a stable clock at each end of the link, but this gives the advantage of noise immunity and eliminates problems of loss of colour sync. which can seriously affect the simple line sequential systems. The AVT mode can operate at four different frame rates of 24, 90, 94 and 188 seconds. The 94- and 188-second modes seem to be favourite. Like Scottie and Martin modes, the AVT picture has a 4:3 aspect ratio.

The AVT system can also operate in a black and white mode which has a frame rate of 125 seconds.

Scan converters

For all-mode operation the Robot 1200C is a popular choice of scan converter. This machine provides pictures in either black and white, or colour with a resolution of 256 pixels by 240 lines. The unit is microprocessor based, and by changing the original software ROMs it can be made to handle Scottie. Martin and AVT pictures. The 1200C can be linked to a home computer to enable pictures to be stored and read from a floppy or hard disk. Earlier versions of the Robot, such as the 400, may be used for 8-second black and white, whilst the 400C provided line sequential colour. Unfortunately, Robot no longer make amateur SSTV equipment.

In Europe a widely used scan converter is the Volker–Wraase SC1, which gives both black and white and high-resolution line sequential colour modes. A newer SC2 version includes the Martin and Scottie modes.

Home computers

Modern 16-bit home computers, such as the Commodore Amiga and the Atari ST, lend themselves quite well to SSTV operation and can be programmed to operate as stand-alone scan converters. The Amiga 500 provides a resolution of 320 × 200 pixels and has a good palette of colours. The newer Amiga 1200 is much better, with a wider range of display modes. For the Amiga an input interface is used to convert the SSTV signals into a stream of 8-bit data which is fed in on a parallel port to the computer. Video frame grabber devices are also available which will allow pictures to be captured from a camera or other video source. The AVT transmission system was developed for use on the Amiga computer, and the software can usually handle the other SSTV modes as well.

The Atari ST can readily be programmed to handle high-resolution black and white pictures using its 320 × 200 low-resolution mode, and it is possible to obtain a 16-level grey scale which gives good pictures. For colour SSTV the ST is limited to low resolution, where it can display 16 colours at a time. By using some form of dithering to produce a wider range of colours it is possible to produce quite good colour SSTV pictures.

The newer Atari Falcon is an ideal machine for SSTV, since its built-in DSP can be used to handle demodulation whilst a 16-bit true colour display mode provides excellent colour pictures. Video digitizers are available to provide input from a video camera or other source, and the built-in DSP can generate the SSTV modulation signal for the transmitter.

In Europe the Sinclair Spectrum has been used for low-resolution black and white SSTV, and although crude in comparison to the 16-bit machines it can provide a cheap entry to the world of SSTV. The IBM compatible PCs can be used for black and white SSTV and are often used in conjunction with the Robot scan converter to provide disk storage for pictures and to generate captions.

The IBM compatible PC can be used for SSTV by adding a decoder board to handle the SSTV signals. The machine must have VGA colour graphics and will usually need around a megabyte of memory.

It seems likely that in the near future some form of digital SSTV could be developed using a modern home computer for encoding and decoding of the picture information. This would probably use a picture data compression system such as JPEG or MPEG1 to reduce the required data rate to perhaps

Figure 7.4 Example of an 8-second monochrome SSTV image received using an Atari Falcon computer

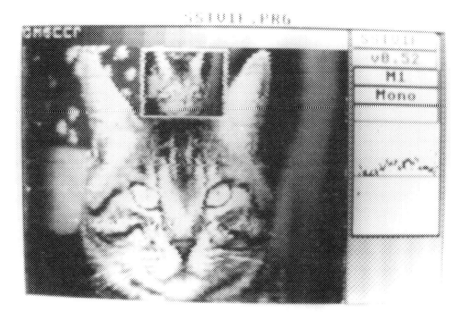

(a)

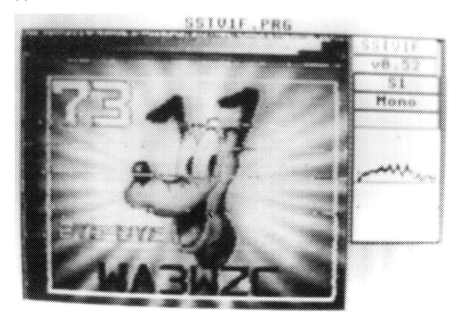

(b)

Figure 7.5 High resolution colour SSTV pictures. (a) Martin mode; (b) Scottie mode

300 bits per second. Data could then be sent using a form of packet transmission. At the receiving end the computer would decompress the data to reproduce the original picture on its screen.

SSTV frequencies
Slow scan signals can be found around the following frequencies in the amateur bands:

3.5 MHz	3735–3745 kHz
7.0 MHz	7035–7045 kHz
14 MHz	14 225–14 235 kHz
	14 170 kHz
21 MHz	21 335–21 345 kHz
28 MHz	28 675–28 775 kHz
144 MHz	144 500 kHz
432 MHz	432 500 kHz

Facsimile (FAX)
Facsimile (FAX) is a method of transmission of pictures as a narrow-band signal which is generally used for weather charts, press photos and amateur picture communication. The system is similar in many ways to SSTV but has a higher picture resolution and a much slower transmission rate. Typically a picture will take about 20 min for transmission. There are no line sync. pulses and the system relies upon the receiver display device having exactly the correct scanning speed. Timing and synchronization signals are sent at the start of the picture transmission and these signals also identify the format of the picture being sent.

At the transmitting end, the picture or chart is wrapped around a rotating drum. A lamp and photocell sensor on a movable head detects the image density as the drum rotates and this effectively scans one line across the chart. After each drum revolution the pickup head is moved axially a short distance along the drum to scan a new line adjacent to the first. This continues until the whole picture or chart has been scanned. At the receiving end a similar drum system is used but in this case the head carries a light source whose intensity can be varied in sympathy with the received signal. The paper used is light sensitive and produces a copy of the original image as the scanning process is carried out. FAX pictures may also be produced by using a personal computer system and a dot matrix, inkjet or laser printer.

Index of cooperation (IOC)
This defines the aspect ratio (width to height) of the picture and is given by:

IOC = drum diameter × line density

Usual values for IOC are 288 and 576.

Line density
This is the number of lines per millimetre on the paper. Usual values are 1.9 lines/mm (IOC = 288) and 3.8 lines/mm (IOC = 576) based on a drum diameter of 152 mm. American machines will usually have the density specified in lines per inch and the drum diameter in inches.

Lines per minute
This is determined by the rotation speed of the scanning drum. The rates normally used are 60, 90, 120 or 240 lines per minute. The drum rotates once for each scan line.

Transmission format
The transmission format is defined by the number of lines per minute and the IOC in use. It is usually shown in the form LPM/IOC. Thus a format of 120/288 means 120 lines per minute with an IOC of 288.

Modulation scheme
On HF, FAX signals use FM (F3C). For black and white charts the frequency shift is:

Black level $f_0 - 400$ Hz
White level $f_0 + 400$ Hz

For photographs, grey levels produce frequencies between these limits.

Reception on the HF bands can be achieved by using the SSB receive mode and offsetting the tuning to give a tone which shifts between 1500 Hz for black and 2300 Hz for white. The FM tone is then demodulated to give the required brightness signal for the recorder. If the upper sideband (USB) receive mode is used, the receiver is tuned 1900 Hz lower than the nominal carrier frequency of the FAX station.

On VHF the FAX signal is converted to a tone which varies from 1500 Hz to 2300 Hz, and this is used to modulate a narrow-band FM transmitter. Weather satellites, such as the NOAA series, use AM with a tone which varies from 1500 to 2300 Hz.

Mode identifier tones
The IOC in use is indicated by a sequence of alternating black and white signals at the start of transmission of a picture. The frequency of the black/white pulses is 300 Hz for an IOC of 576 and 675 Hz for an IOC of 288. On receipt of these signals the FAX machine automatically selects the required IOC.

Phasing signal
At the start of picture transmission an alternating black and white signal is used to define the start point of the scanning lines so that the receiving

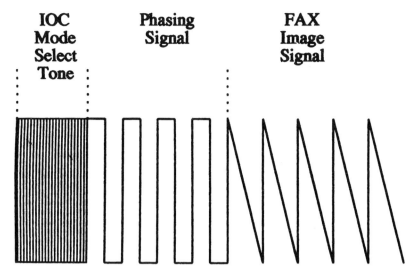

IOC
Mode
Select
Tone

Phasing
Signal

FAX
Image
Signal

Figure 7.6 The phasing and IOC tone signals which are sent at the start of a FAX picture

scanner can synchronize to the received signal. One black/white cycle occurs per scan line with half the line black and half white. Some systems use a phasing signal which is 95% black with a 5% white pulse. The start of the line is indicated by the start of the white part of the phasing signal.

Transmission time

The time taken to send a complete picture depends upon the combination of the number of lines per minute, the IOC and the physical size of the picture being transmitted. Typical transmission times for weather charts are:

Lines/min IOC		Time (min)
60	288	18.8
90	288	12.5
90	576	25.0
120	288	9.4
120	576	18.8
240	288	4.7
240	576	9.4

Amateur FAX

Amateur FAX generally uses the 120/288 format and may sometimes be found on HF around the following frequencies:

3600 kHz
7040 kHz
14 100 kHz (Europe)
14 245 kHz (USA)

There is a European FAX net which meets on 14 100 kHz most days at around 1700 GMT, and a news broadcast is made on Saturdays on 3600 kHz at 1700 GMT and on Sundays at 1000 GMT on about 14 102 kHz.

FAX equipment

It is possible to obtain used FAX machines and use these for transmission and reception of FAX images. Note that the standard office FAX machine uses a different system and is not suitable for HF FAX transmission or reception.

The alternative approach is to use a computer to decode and display received FAX images and to generate and transmit a FAX image. Many

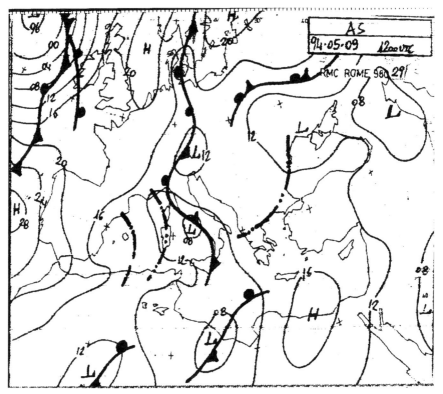

Figure 7.7 Example of a FAX image produced using an Atari ST computer and a dot matrix printer

systems use an external interface which decodes the tones and provides the accurate timing clock required. In this case the computer generates the display or provides the data for an image to be transmitted. It is possible on some computers, such as the Commodore Amiga and the Atari ST or Falcon, to decode the received FAX signal by software so that only a simple signal clipper is needed for the input interface. Once the FAX image has been stored in the computer's memory it can either be displayed on the TV or monitor screen or printed out using a dot matrix or laser printer.

Some versions of the AEA PK232 packet TNC can be used to receive FAX pictures if the appropriate software is used with the associated computer.

8

Utility stations

Apart from broadcasting and amateur radio stations, most of the signals on the HF, VHF and UHF bands provide special services and are generally referred to as utility stations. Many of these provide communications for government, commercial and military purposes using a variety of transmission modes. Some of these signals will be scrambled or encrypted to provide privacy and security.

It should be noted that these stations are not intended for public broadcasting and their reception is not covered by the normal radio receiving licences. Listeners are reminded that the divulgence or improper use of any information gained from these transmissions constitutes a violation of the international telecommunications laws and is a criminal offence in most countries.

Utility stations can provide useful test signals for checking RTTY and FAX equipment since, unlike amateur stations, they usually operate to some form of schedule and transmit for long periods. The signals from these stations are also a useful guide to propagation conditions, since in general they produce stronger and more consistent signals than amateur stations.

Press RTTY stations

A number of press agency stations broadcast news on the HF bands. Most of these use the 50-baud RTTY mode of transmission with the ITU No. 2 alphabet. Frequency shift is usually 425 Hz, with the higher frequency being a 'mark'. A few stations use 75- or 100-baud transmissions and some operate with reversed frequency shift. Transmissions may be in a variety of languages, such as English, French, German and Spanish. Transmissions in Russian and Arabic use special versions of the ITU code and are not readily deciphered. The bands where these stations are to be found are:

5200–5500 kHz
5800–6000 kHz
6900–7000 kHz
7400–8200 kHz
8900–9500 kHz
9800–10 000 kHz
10 500–11 200 kHz
11 400–11 600 kHz
13 400–13 800 kHz
14 350–15 000 kHz
15 500–16 500 kHz
18 200–18 700 kHz
20 200–20 500 kHz

Some typical frequencies for press agency RTTY stations (in kilohertz) are:

MAP, Rabat, Morrocco:	7842.4, 10 213, 10 634, 14 769, 15 752, 16 000
TANJUG, Belgrade, Serbia:	5240, 7592, 7658, 7806, 7996, 12 212, 13 440
XINHUA, Beijing, China:	10 982, 11 133, 12 265, 14 367
ATA, Tirana, Albania:	7850, 9133, 9430
DyN, Buenos Aires:	4549.5, 7954.5
NA, Buenos Aires:	3840, 10 805
GNA, Bahrein:	14 764 (1500–1700 GMT)
JANA, Tripoli, Libya:	12 186, 15 462
ROMPRES, Rumania:	6972, 9797, 12 110
CNA, Taiwan:	7695, 10 235, 10 960, 13 673
TAP, Tunis, Tunisia:	13 610, 14 800

Transmission times vary, but most European, African and Asian stations are likely to be heard in the morning or afternoon. South American stations tend to be audible in Europe late at night and in the early hours of the morning. Stations usually run an unmodulated carrier between news items.

Press FAX pictures
Although most press photographs are now transmitted via communications satellite channels or by landline, a few stations still transmit these pictures on the HF bands using the photo FAX system. The format used is generally 60/288, with a frequency shift of 400 Hz each side of the carrier frequency. The polarity of the video signal and scan direction are usually reversed to produce a negative image. The computer software used should be able to cope with this. The direction of scanning for press photos is from right to left. The program must also be able to display at least eight shades of grey, and preferably more, to give an acceptable picture.

Station JJC in Tokyo and KYODO Singapore transmits a newspaper in Japanese text and also weather maps using normal scan 60 LPM FAX.

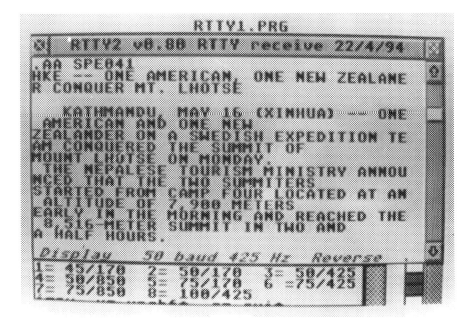

Figure 8.1 Typical news item from an RTTY press station on the HF bands

Figure 8.2 Typical FAX picture received from a station in Argentina on the HF bands

Pictures from this station can be received using a basic computer program designed for receiving weather charts.

Some frequencies in use for pictures are:

DyN, Buenos Aires:	8167.5, 9242 kHz. Press pictures 60/288 (2100–0200 GMT).
JJC, Tokyo:	8467.5, 12 745.5, 16 971 kHz. News text and weather charts 60/288.
DPA, Germany:	139 kHz. Press pictures 120 LPM from AP (Associated Press).
KCNA, Korea:	11 475.7 kHz. Press pictures 60/288 (2300 GMT).
KYODO News, Singapore:	16 270 kHz. News text and weather maps 60/288.

Meteorological RTTY stations

One major group of utility stations are those which provide weather and other meteorological information. These are normally government-operated stations.

Figure 8.3 Typical synoptic weather report received via RTTY on the HF bands

Many of the meteorological stations transmit weather information using RTTY at 50 baud. Unlike amateur and press teletype stations, these signals use a wide shift of 850 kHz. The transmissions are coded and appear as strings of four- or five-figure groups giving data on a variety of meteorological measurements. These are known as synoptic codes and it is possible to obtain a decoder unit which will translate the codes into plain English data, and these can be displayed or printed by using an IBM compatible PC. For more information on the meaning of the codes used, readers are referred to the *Meteo Code Manual* published by Klingenfuss.

Typical frequencies (kilohertz) are:

Bracknell, UK	4489, 6835, 10 551, 14 356
Shannon, Eire	8145, 11 440
Paris, France	7642.7, 13 593.8
Moscow, Russia	5020, 5140, 7685, 7855, 9190, 10 830, 11 450
Beijing, China	7350, 9195, 10 320, 14 340
Halifax, Canada	6496, 10 536, 13 510
Boston, USA	8416, 12 579
Hamburg	4583, 7646, 11 039, 11 638
Tokyo, Japan	7402, 14 880

FAX weather charts

Many meterological stations use FAX transmissions to broadcast weather charts of various kinds. Most use either 120/288 or 120/576 formats, although Russian stations often use 60/288, 90/288 and 90/576 formats as well. The charts are simple black and white line drawings but some stations also transmit weather satellite photographs at certain times during the day.

Typical frequencies (kilohertz) are:

Bracknell, UK	2619.5, 3289.6, 4610, 4782, 8040, 9203, 11 086, 14 436
RN, London	4307, 6446, 8331.5, 12 844.5
Melbourne, Australia	11 030, 13 920
Rome, Italy	4777.5, 8146.6, 13 597.4
Madrid, Spain	6918.6, 10 250
Halifax, Canada	6496.4, 10 536, 13 610
Beijing, China	8122, 10 117, 14 367, 16 025
Moscow, Russia	5150, 5355, 6880, 7670, 7750, 10 230, 10 710, 11 525, 13 470
Hamburg, Germany	3855, 7880
USN, Norfolk Va, USA	9318, 10 865, 20 015

The US Navy station NAM at Norfolk Virginia transmits GOES satellite pictures at 0200, 0515, 0545, 1115, 1400, 1715, 1745 and 1945 GMT.

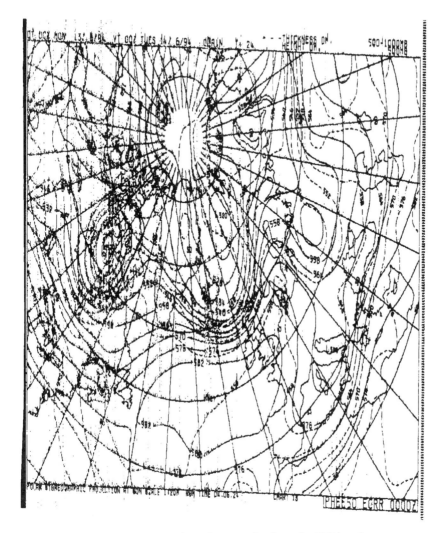

Figure 8.4 A typical FAX weather chart received on the HF bands

The aircraft bands

Several bands are set aside for use by aircraft and ground control stations. Most of the voice transmissions between aircraft and ground stations in these bands are for air traffic control. The HF bands are used for major transoceanic routes, whilst VHF and UHF are used for communications along the airways routes over land or for approach and landing. The other stations to be heard on the HF bands are the VOLMET stations, which give weather conditions for various airports.

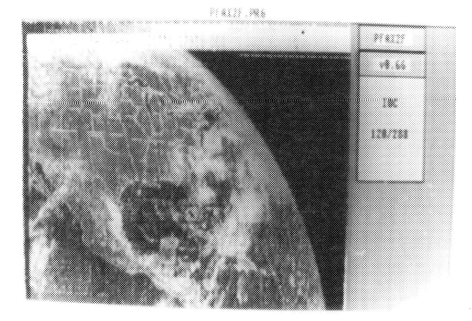

Figure 8.5 A satellite picture received by FAX from the US Navy Metcentre in Norfolk Va

HF aircraft communications
The bands used for oceanic communications on HF (kilohertz) are:

5480–5730
6525–6775
8800–9040
10 005–10 100
11 175–11 400
13 200–13 360
17 900–18 030
21 924–22 000

For the main intercontinental air routes the world is divided up into a number of air traffic control areas which cover North Atlantic, South Atlantic, Caribbean, Africa, Middle East, Indian Ocean, Far East, South America, North Pacific, South Pacific and Europe.

Some commonly used frequencies (kilohertz) for the various major intercontinental routes are:

North Atlantic   3419, 5598, 5616, 5649, 8825, 8891, 11 279, 11 336, 13 306
Caribbean   5520, 5550, 6577, 8846, 8918, 11 387
Middle East   3467, 5667, 8918, 13 288, 13 312

Africa	5493, 5652, 8861, 8895, 11 300, 11 360, 13 279
Indian Ocean	5634, 13 291, 13 306, 17 961
Far East	5655, 5658, 6556, 6571, 8942, 13 309, 13 318

Some channels in the HF aircraft bands are used by individual airlines as a company frequency whilst others are allocated to major airports for long-range communications. Voice transmissions in the aircraft bands generally use SSB (upper sideband) and channels are normally spaced at 3-kHz intervals, although adjacent channels are rarely used in the same control area. Some transmissions are made using CW or RTTY modes.

Commercial and military aircraft normally carry transponders which provide identification for ground radar stations. These are referred to as 'squawk' signals. Selective calls (Selcalls) consist of a combination of audio tones which are transmitted to the aircraft, decoded by its receiver and used to sound an alarm. Each aircraft has its own combination of tones so that one particular aircraft can be alerted when the controller wishes to communicate with it. This allows the pilot or engineer to concentrate on other tasks without having to monitor the radio channel.

Callsigns used by commercial airliners are usually based on the airline name and flight number. Private aircraft identify by using the aircraft registration letters. Military aircraft use callsigns allocated for their particular sortie or operation.

VHF and UHF aircraft communications

Most local air traffic control, including approach, landing and communications whilst the aircraft is on the ground, is carried out using the VHF or UHF bands.

Civil aircraft and airports use frequencies in the band 118.00–136.00 MHz. Transmissions use AM (A3E) and channels are spaced at 50-kHz intervals. Simplex operation is used with the aircraft and tower on the same frequency.

The easiest channels to find are the en route control channels for the major control zones. These control the aircraft from the international routes to the radar or approach operators at the main airports. Some typical frequencies (MHz) in use are:

London ATCC	125.800, 125.950, 126.450, 126.825, 128.400, 132.050, 132.600, 134.450
London Military	124.600, 124.750, 128.700, 134.700, 135.150, 135.275
Scottish ATCC	124.050, 124.900, 126.250, 128.500, 131.300, 133.200, 135.850

Approach control may be heard if you are close to the airport. Some frequencies (mHz) in use are:

Heathrow	119.200, 119.500, 120.400, 127.550
Gatwick	125.875, 134.225

Stansted	125.550
Manchester	119.400
Birmingham	131.325

Military airfields and aircraft generally use the band 225.00–400.00 MHz, although some frequencies in the 118–136-MHz band are also used. VHF airband transmissions are simplex using AM (A3F).

VOLMET transmissions
VOLMET stations provide weather and meteorological broadcast services for aircraft. They generally operate all day but switch frequency from time to time as propagation changes.

VOLMET stations frequencies (kHz) for the various world areas are:

Europe
Shannon	3413, 5505, 8957, 13 264
RAF, London	4722, 11 200

North Atlantic
New York	3485, 6604, 10 051, 13 270
Gander, Canada	3485, 6604, 10 051, 13 270

Africa
Brazzaville	10 057, 13 261

South East Asia
Bangkok	6676, 11 387
Singapore	6676, 11 387
Sydney	6676, 11 387

Pacific
Auckland, New Zealand	8828, 13 282
Hong Kong	8828, 13 282
Hawaii	8828, 13 282

Stations sharing the same frequency work on a timeshare basis with each station making two reports per hour. These HF VOLMET stations use SSB transmissions. VHF VOLMET stations in the UK operate on 126.6 and 128.6 MHz, using AM (A3E) transmissions.

Air band terminology
Some common terms used on the aircraft bands are:

CAV OK	Ceiling and visibility OK
FIR	Flight information region
QNH	Barometric pressure at sea level
QFE	Barometric pressure at an airport
NO SIG	No significant weather conditions

CAT   Clear air turbulence
DME   Distance-measuring equipment
IFR   Instrument flight rules
RVR   Runway visual range
Stratus   Low misty cloud
VFR   Visual flight rules

HF maritime radio

A large part of the HF radio spectrum is allocated for use by ships and other maritime services. There are many coastal stations which provide communication with ships. The VHF band is used for short-range communications such as control of traffic in port areas and for private boats operating near the coast.

The HF maritime bands (kHz) are:

4000.0–4438.0
6200.0–6525.0
8100.0–8812.0
12 230.0–13 200.0
16 360.0–17 410.0
22 000.0–22 855.0

Maritime HF Morse (CW) channels

With the advent of modern radio equipment many ships today do not carry a radio operator, so that radio communication is often carried out by a navigating officer. As a result most transmissions use either telephony or RTTY and the use of Morse code is steadily declining.

For CW (Morse) operation the channel spacing depends upon the frequency band used. Different groups of channels are used for calling and working frequencies.

CW calling channels are:

4180.0–4187.2 kHz (18 ch. 400 Hz apart)
6270.0–6280.8 kHz (18 ch. 600 Hz apart)
8360.0–8374.4 kHz (18 ch. 800 Hz apart)
12 540.6–12 561.0 kHz (18 ch. 1.2 kHz apart)
16 720.0–16 748.8 kHz (18 ch. 1.6 kHz apart)
22 228.0–22 246.0 kHz (10 ch. 2.0 kHz apart)

CW working channels are:

4188.5–4219.0 kHz (62 ch. × 500 Hz)
6282.75–6325.0 kHz (44 ch. × 750 Hz)
8377.0–8435.0 kHz (116 ch. × 500 Hz)
12 565.5–12 652.0 kHz (174 ch. × 500 Hz)

16 754.0–16 858.5 kHz (208 ch. × 500 Hz)
22 250.5–22 310.0 kHz (120 ch. × 500 Hz)

Maritime HF telephony (SSB) channels

For voice communications between ship and shore, the signal sideband (USB) mode is used and in most cases duplex operation, where ship and shore stations are on different frequencies, is used. The channel arrangement is as follows:

4-MHz band 27 channels at 3-kHz spacing

Channel	Ship	Shore
401	4065 kHz	4357 kHz
427	4143 kHz	4435 kHz

6-MHz band, 8 channels at 3-kHz spacing

Channel	Ship	Shore
601	6200 kHz	6501 kHz
608	6221 kHz	6522 kHz

8-MHz band, 32 channels at 3-kHz spacing

Channel	Ship	Shore
801	8195 kHz	8719 kHz
802	8198 kHz	8722 kHz
832	8288 kHz	8812 kHz

12-MHz band, 41 channels at 3-kHz spacing

Channel	Ship	Shore
1201	12 230 kHz	13 077 kHz
1202	12 233 kHz	13 080 kHz
1241	12 350 kHz	13 197 kHz

16-MHz band, 56 channels at 3-kHz spacing

Channel	Ship	Shore
1601	16 360 kHz	17 242 kHz
1602	16 363 kHz	17 245 kHz
1656	16 525 kHz	17 497 kHz

18-MHz band, 15 channels at 3-kHz spacing

Channel	Ship	Shore
1801	18 780 kHz	19 755 kHz
1802	18 791 kHz	19 758 kHz
1815	18 822 kHz	19 787 kHz

22-MHz band, 53 channels at 3-kHz spacing

Channel	Ship	Shore
2201	22 000 kHz	22 696 kHz
2202	22 003 kHz	22 699 kHz
2253	22 156 kHz	22 852 kHz

25-MHz band, 10 channels at 3-kHz spacing

Channel	Ship	Shore
2501	25 070 kHz	26 145 kHz
2502	25 073 kHz	26 148 kHz
2510	25 097 kHz	26 172 kHz

Maritime HF RTTY channels
Nearly all maritime RTTY transmissions today use the SITOR system for transmission with a frequency shift of 170 Hz. A few Russian stations use Baudot RTTY at 50 baud. Operation is usually duplex, with the ship station on one frequency and the shore station on another.

4-MHz band, 18 channels at 500 Hz spacing

Channel	Ship	Shore
1	4172.5 kHz	4210.5 kHz
2	4173.0 kHz	4211.0 kHz
10	4177.0 kHz	4215.0 kHz
12	4178.0 kHz	4215.5 kHz
19	4181.5 kHz	4219.0 kHz

Channel 11, 4177.5 kHz, is a ship distress frequency.

6-MHz band, 25 channels at 500-Hz spacing

Channel	Ship	Shore
1	6263.0 kHz	6314.5 kHz
2	6263.5 kHz	6315.0 kHz

10	6267.5 kHz	6319.0 kHz
12	6268.5 kHz	6319.5 kHz
26	6275.5 kHz	6326.5 kHz

Channel 11, 6268.0 kHz, is a ship distress frequency.

8-MHz band, 39 channels at 500-Hz spacing

Channel	Ship	Shore
2	8377.0 kHz	8417.0 kHz
3	8377.5 kHz	8417.5 kHz
40	8396.0 kHz	8436.0 kHz

Channel 1, 8376.5 kHz, is a ship distress frequency.

12-MHz band, 145 channels at 500-Hz spacing

Channel	Ship	Shore
1	12 477.0 kHz	12 597.5 kHz
2	12 477.5 kHz	12 598.0 kHz
86	12 519.5 kHz	12 622.0 kHz
88	12 520.5 kHz	12 622.5 kHz
146	12 549.5 kHz	12 651.5 kHz

Channel 87, 12 520 kHz, is a ship distress frequency.

16-MHz band, 192 channels at 500-Hz spacing

Channel	Ship	Shore
1	16 683.5 kHz	16 807.0 kHz
2	16 684.0 kHz	16 807.5 kHz
23	16 694.5 kHz	16 818.0 kHz
25	16 695.0 kHz	16 818.5 kHz
101	16 733.5 kHz	16 856.5 kHz
102	16 739.0 kHz	16 857.0 kHz
193	16 784.5 kHz	16 902.5 kHz

Channel 24, 16 695.0 kHz, is a ship distress frequency.

22-MHz band, 135 channels at 500-Hz spacing

Channel	Ship	Shore
1	22 284.5 kHz	22 376.5 kHz
2	22 285.0 kHz	22 377.0 kHz
134	22 351.0 kHz	22 443.0 kHz
135	22 351.5 kHz	22 443.5 kHz

25-MHz band, 40 channels at 500-Hz spacing

Channel	Ship	Shore
1	25 173.0 kHz	26 101.0 kHz
2	15 173.5 kHz	26 101.5 kHz
39	25 192.0 kHz	26 120.0 kHz
40	25 192.5 kHz	26 120.5 kHz

Most of the other frequencies in the shipping bands are used by coastal stations and for special facilities such as selective calling.

VHF maritime radio

For local working near the shore and in harbour the VHF maritime band is used. The VHF band from 156 to 157.5 MHz was originally divided into 28 channels spaced 50 kHz apart for use by ships, with a corresponding set of duplex channels from 160.5 to 162 MHz for coastal stations. These channels are numbered 01 to 28.

Channel	Ship frequency (MHz)	Shore frequency (MHz)
01	156.050	160.650
02	156.100	160.700
03	156.150	160.750
04	156.200	160.800
05	156.250	160.850
06	156.300	160.900
07	156.350	160.950
08	156.400	161.000
09	156.450	161.050
10	156.500	161.100
11	156.550	161.150
12	156.600	161.200
13	156.650	161.250
14	156.700	161.300
15	156.750	161.350
16	156.800	161.400
17	156.850	161.450
18	156.900	161.500
19	156.950	161.550
20	157.000	161.600
21	157.050	161.650
22	157.100	161.700
23	157.150	161.750
24	157.200	161.800
25	157.250	161.850
26	157.300	161.900
27	157.350	161.950
28	157.400	162.000

To provide more channels, a further set of 28 channels was added by inter-leaving them with the first set. These new channels were numbered from 60 to 88.

Channel	Ship frequency (MHz)	Shore frequency (MHz)
60	156.025	160.625
61	156.075	160.675
62	156.125	160.725
63	156.175	160.775
64	156.225	160.825
65	156.275	160.875
66	156.325	160.925
67	156.375	160.975
68	156.425	161.025
69	156.475	161.075
70	156.525	161.125
71	156.575	161.175
72	156.625	161.225
73	156.675	161.275
74	156.725	161.325
75	156.775	161.375
76	156.825	161.425
77	156.875	161.475
78	156.925	161.525
79	156.975	161.575
80	157.025	161.625
81	157.075	161.675
82	157.125	161.725
83	157.175	161.775
84	157.225	161.825
85	157.275	161.875
86	157.325	161.925
87	157.375	161.975
88	157.425	162.025

Coastguards operate on 156.000 MHz. Distress and initial calls are made using channel 16. Intership calls and the coastguards use channel 06. For port control and operations, channels 19 to 22 and channels 79 and 80 are used.

Other utility stations

The HF bands contain many other utility stations, including telecommunications links, such as telephone or telex services, and military communications. These are often difficult to identify, or use transmission modes which are not readily decoded. Several books are available which provide lists of these confidential frequencies and give some indication of times at which signals may be heard.

The VHF and UHF bands are used for public service communications systems, such as police, fire or ambulance services, and private mobile radio systems such as those used by taxis. Most of these transmissions have the base station and the mobile transmitter on different frequencies. In some cases base stations are operated as repeaters so that mobile stations can talk to one another. Some police radio networks use encryption to prevent unauthorized listening.

Other stations in these bands include the cellular radio links for mobile car telephones and some paging services. Two small bands are used for radio-controlled models.

Private mobile radio (PMR)
Frequency allocations (MHz) are:

85.0125–86.2875	Base Tx, two-frequency simplex
71.3125–72.7875	Mobile Tx, 12.5-kHz channels
86.9625–87.5000	Base Tx, two-frequency simplex
76.9625–77.5000	Mobile Tx, 12.5-kHz channels
86.3000–86.7000	Simplex, 12.5-kHz channels
165.050–168.250	Base Tx, two-frequency simplex
169.850–173.050	Mobile Tx, 12.5-kHz channels
168.950–169.850	Simplex, 12.5-kHz channels
176.500–183.500	Base Tx, split simplex
184.500–191.500	Mobile Tx, 12.5-kHz channels
200.500–207.500	Base Tx, split simplex
192.500–199.500	Mobile Tx, 12.5-kHz channels
208.500–215.500	Base Tx, split simplex
216.500–223.500	Mobile Tx, 12.5-kHz channels
453.025–453.975	Base Tx, split simplex
459.325–460.475	Mobile Tx, 12.5-kHz channels
456.000–456.975	Base Tx, split simplex
461.500–462.475	Mobile Tx, 12.5-kHz channels
448.006–448.994	Base Tx (London area only)
431.006–431.994	Mobile Tx (London area only)
163.0375–164.4250	Base Tx, radiophone
158.5350–159.9125	Mobile Tx, split simplex
935.0125–949.9875	Base Tx, cellular radio
890.0125–904.9875	Mobile Tx, duplex

Emergency services
Emergency services such as police, fire and ambulance operate in the following bands of frequencies (MHz):

143.000–144.000
146.000–148.000
152.000–153.000
154.000–156.000
451.000–453.000
465.000–467.000

The duplex frequencies for 143 MHz are at 152 MHz, and those for 148 MHz are at 156 MHz. Stations may operate duplex with a variety of offsets, depending upon the area. Channels are normally spaced 25 kHz apart. Most stations use FM but some in the 154–156 MHz band use AM. Some police channels use scrambling or encryption techniques to prevent unauthorized access.

In most cases only the base station will be heard unless the mobile is operating nearby. For some services the base station acts as a repeater so that all users of the channel can hear both base and mobile stations.

Many of the mobile frequencies are used for taxi services, buses and general business communications. Some channels are used for data transmission. For most channels activity is infrequent so that a casual scan through the band may seem to indicate that there is no activity.

Radio paging transmitters
Frequencies (MHz):

26.9570–27.2830	On-site paging
31.7250–31.7750	On-site paging
153.025–153.475	Wide-area paging
161.000–161.100	Acknowledgement channels
454.0125–454.8250	Wide-area paging
459.100–459.500	On-site paging

Model control bands
Frequencies (MHz):

35.005–35.205
458.50–459.50

Citizens band radio
In the UK two bands are allocated for CB radio operation and on both bands FM is used. The lower band is split into two sections each containing 40 channels. The original UK CB channels occupy the frequencies from 27.60125 to 27.99125 MHz. The second section of the lower band, which has recently been released for use in the UK, contains the 40 CEPT European channels and ranges from 26.965 to 27.405 MHz.

A licence for the use of a CB receiver or transmitting in the UK can be obtained from Subscription Services in Bristol.

UK 27-MHz CB channels

Channel	Frequency (MHz)
1	27.60125
2	27.61125
3	27.62125
4	27.63125

5	27.64125
6	27.65125
7	27.66125
8	27.67125
9	27.68125
10	27.69125
11	27.70125
12	27.71125
13	27.72125
14	27.73125
15	27.74125
16	27.75125
17	27.76125
18	27.77125
19	27.78125
20	27.79125
21	27.80125
22	27.81125
23	27.82125
24	27.83125
25	27.84125
26	27.85125
27	27.86125
28	27.87125
29	27.88125
30	27.89125
31	27.90125
32	27.91125
33	27.92125
34	27.93125
35	27.94125
36	27.95125
37	27.96125
38	27.97125
39	27.98125
40	27.99125

Channel 9 is reserved as an emergency channel and is monitored continuously by a series of volunteer stations. This is used for road traffic accidents and similar incidents where emergency help is required.

Channel 14 is used as a calling channel on which communication is first established before moving to another channel. Channel 19 is used as a calling channel by truck drivers, following the practice in the USA where channel 19 is used as the calling channel.

Specifications for 27-MHz equipment are:

Maximum output power: 4 W
Modulation: FM (F3E), 2.5-kHz deviation
Antenna: Single vertical wire or rod
Antenna length: 1.5 m maximum, including loading coil

All equipment for use on the 27-MHz band must be built to a strict specification and carry an authorization label. The manufacture, import, use and even possession of AM- or SSB-type CB equipment is illegal in the UK.

27-MHz CB in Europe
The CEPT (Committee of European Posts and Telecommunications) has designated a set of 40 channels for CB radio throughout Europe. These channels are now available to UK users in addition to the 40 UK CB channels listed above.

Channel	Frequency (MHz)
1	26.965
2	26.975
3	26.985
4	26.995
5	27.005
6	27.015
7	27.025
8	27.035
9	27.045
10	27.055
11	27.065
12	27.075
13	27.085
14	27.095
15	27.105
16	17.115
17	27.125
18	27.135
19	27.145
20	27.155
21	27.215
22	27.225
23	27.235
24	27.245
25	27.255
26	27.265
27	27.275
28	27.285
29	27.295
30	27.305
31	27.315
32	27.325
33	27.335
34	27.345
35	27.355
36	27.365
37	27.375
38	27.385
39	27.395
40	27.405

CB radio in North America
During the high part of the sunspot cycle CB signals from the USA are often heard in Europe. These transmissions use AM (A3E) or SSB (J3E). There are 80 channels available and the frequencies are the same as for the

European channels, but in the USA the upper 40 channels are numbered from 41 to 80.

UK 934-MHz personal radio

The UHF band from 934.025 MHz to 934.995 MHz was allocated for personal radio in the UK at the same time that the 27-MHz CB was authorized. Twenty channels were allocated at 50-kHz spacing as follows:

Channel	Frequency (MHz)
1	934.025
2	934.075
3	934.125
4	934.175
5	934.225
6	934.275
7	934.325
8	934.375
9	934.425
10	934.475
11	934.525
12	934.575
13	934.625
14	934.675
15	934.725
16	934.775
17	934.825
18	934.875
19	934.925
20	934.975

Other specifications for equipment on this band are:

Maximum output power: 8 W
Modulation: FM (F3E) 5-kHz deviation
Antenna: Up to 12-element beam permitted
Vertical polarization only
Only speech communication is permitted
Channels 10 and 20 are generally used as calling channels

As for 27 MHz, all equipment for use on this band must carry an authorization label to indicate that it meets the required specifications.

In the late 1980s this band was reallocated for use by a digital communications system and permission to manufacture or import 934-MHz equipment was withdrawn. The existing users of the band are, however, permitted to continue operation until the new digital system becomes operational. As a result of this decision, activity on this band has decreased.

9

Space communications

This section covers the transmission and reception of signals from space vehicles, via communications satellites and via reflection from other bodies in space such as the moon.

Space satellites can generally be grouped into three major types according to the type of orbit they follow as they travel over the surface of the earth. The three orbit types are low-altitude near-circular orbits, high-altitude elliptical orbits and synchronous or geostationary orbits. For tracking calculations the low-orbit satellites are usually assumed to have circular, or near-circular, orbits around the earth as shown in Figure 9.1.

Satellite orbital parameters

For circular- and elliptical-orbit satellites, the position of the satellite relative to the ground station changes with time, and communication is only

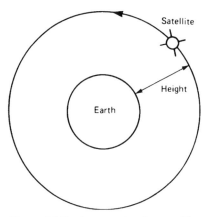

Figure 9.1 Basic near-circular satellite orbit

possible when the satellite is in view of the ground station. For low-altitude satellites the orbit is usually assumed to be circular to simplify tracking calculations. The set of parameters required for calculating the position of the satellite relative to the ground station is usually referred to as a set of Keplerian elements.

The main parameters for a low-altitude near-circular orbit satellite are:

Apogee   The point in its orbit where the satellite is furthest from the earth.

Perigee   The point in the orbit where the satellite is closest to earth.

Ascending node (EQX)   The point in the orbit where the satellite crosses the earth's equatorial plane travelling south to north.

Descending node   The point of equatorial crossing when travelling north to south.

EQX time   The time at which EQX occurs given in universal time code (utc).

EQX longitude   The position around the equator at which EQX occurs. Measured as degrees west.

Period   The time taken for one complete orbit (i.e. between successive EQX times).

Inclination   The angle between the orbital plane of the satellite and the equatorial plane of the earth.

AOS (acquisition of signal)   The time at which the satellite comes above the horizon and in sight of the ground station.

LOS (loss of signal)   The time at which the satellite drops below the horizon out of sight of the ground station.

Azimuth   The direction of the satellite in the horizontal plane relative to the ground station. Measured as an angle clockwise from north.

Elevation   The angle of the satellite above the horizontal relative to the ground station.

Russian satellites have inclination angles less than 90° and the satellite orbit climbs to the east of the north pole on the ascending part of the orbit as shown in Figure 9.2. American satellites have inclination angles greater than 90° and their orbits climb to the west of the north pole.

Low-orbit satellites give relatively short communications periods when the satellite is visible from the ground station.

For satellites which are in an elliptical orbit the apogee is very much greater than perigee, as shown in Figure 9.3. When the satellite is near apogee the period of visibility to the ground station is much longer than for a low-orbit satellite. A set of Keplerian elements is used to calculate the satellite track.

Additional parameters used for tracking these satellites are:

Eccentricity   A value which indicates the flatness of the ellipse relative to a circle which has an eccentricity of zero.

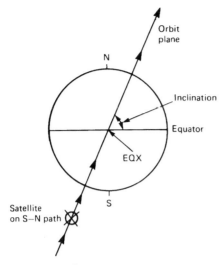

Figure 9.2 Diagram showing the inclination and EQX of a satellite orbit

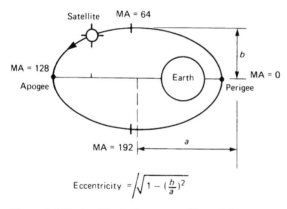

$$\text{Eccentricity} = \sqrt{1 - \left(\frac{b}{a}\right)^2}$$

Figure 9.3 Basic elliptical-type satellite orbit

Right ascension of the ascending node (RAAN)   This defines the satellite position in relation to a set of coordinates based on a fixed point in space. This reference is the first point of Aries and the RAAN is the angle between the line of nodes and a line to the point in Aries.

Argument of perigee   The angle between a line from the earth centre to perigee and a second line from the earth centre to the point where the orbit crosses the equator at the EQX time.

Mean anomaly   The angular position (phase) of the satellite around its orbit measured from perigee.

Mean motion   The number of passes through the perigee point per day.

Epoch The time at which the parameters were measured. Usually given as year, day and decimal fraction of day.

Calculations for predicting the AOS and LOS times, azimuth and elevation angles are relatively complex and require the use of an electronic calculator or personal computer. Satellite tracking programs for use on popular home computers are available from a number of sources.

Most computer programs for satellite orbit predictions can be used to calculate times at whch the satellite is visible for both circular and elliptical orbits. Some programs show a map of the world and display the position of the satellite relative to local time. Alternatively the program may be run in real time, showing the current satellite position. The program can also be made to print out times and elevations for any passes where the satellite is visible from a specified location. Today, Keplerian element sets are available for both elliptical- and near-circular-oribit satellites, so it is convenient to use the same prediction program for all of them. The computer can hold data for all of the satellites of interest, so if you tell the program which satellite you want and the time period for the prediction the program will print out details of any orbital passes where communication is possible.

Orbital parameter newscasts
Some news broadcasts and nets on the amateur bands provide information on the orbit parameters of amateur satellites as follows:

ARRL News: EQX times and longitudes
Mon–Fri
CW at 0100, 0400, 1500 and 2200 GMT on 3580, 7080, 14 070, 21 080, 28 080 kHz
SSB at 0230 and 0530 GMT on 3990, 7290, 14 290, 21 390, 28 590 kHz
RTTY at 0200, 0500, 1600 and 2300 GMT on 3625, 7095, 14 095, 21 095, 28 095 kHz

AMSAT (USA) Net: Parameters and other satellite news
Wednesday 0200 GMT on 3855 kHz SSB
Sunday 1800 GMT on 21 280 kHz SSB
Sunday 1900 GMT on 14 282 kHz SSB

AMSAT (Europe) Net: Parameters and satellite news
Saturday 1000 GMT on 14 280 kHz SSB

AMSAT (UK) Net: Satellite news
Sunday 1015 local time on 3780 kHz SSB

Other sources of satellite parameters are the magazines of AMSAT, and many packet radio bulletin boards also have latest details of satellites and the latest Keplerian element sets.

Geostationary satellites

If the satellite is placed in an equatorial orbit at a height of 22 247 miles, the satellite travels around its orbit in 24 hours. If the satellite motion is in the same direction as the rotation of the earth, the satellite remains at a fixed position above the equator. This type of orbit is known as synchronous or geostationary.

The geostationary satellite orbit is called the Clarke belt after Arthur C. Clarke, who proposed the idea of geostationary satellites in the early 1940s. The Clarke belt lies over the equator and is divided into a series of orbital slots at intervals of about 3°. Each orbital slot is defined by the longitude line over which the satellite is placed.

The elevation and azimuth angles for the ground receiving antenna can be calculated from the satellite longitude position and the latitude and longitude of the ground station. Only those satellites which appear above the horizon when viewed from the ground station can be received. Reception will also depend upon the direction in which the satellite antenna is pointed. The area on the earth's surface covered by the beam of the satellite antenna is generally known as its 'footprint'.

Geostationary satellites are used for the main intercontinental communications links and for TV signal distribution and broadcasting. Currently there is no amateur radio activity from geostationary satellites.

Amateur satellite bands

A number of amateur radio satallites for communications and for scientific experiments have been placed into orbit in space. The following segments of the amateur bands are allocated for use with space satellites:

7.000–7.100 MHz
14.000–14.250 MHz
21.000–21.450 MHz
28.000–29.700 MHz
144.00–146.00 MHz
435.00–438.00 MHz
1260.0–1270.0 MHz   Uplink only
2400.0–2450.0 MHz
5650.0–5670.0 MHz   Uplink only
5830.0–5850.0 MHz   Downlink only
10.450–10.500 GHz
24.000–24.050 GHz
47.000–47.200 GHz
75.500–76.000 GHz
142.00–144.00 GHz
248.00–250.00 GHz

Satellite communications transponders

On satellites used for communications a wideband transponder is used which accepts all signals within the uplink passband and retransmits them at corresponding frequencies within the downlink passband. The uplink and downlink are always on different bands. Transponders may be either non-inverting or inverting types. An inverting transponder generally gives less trouble with Doppler shift, since the uplink and downlink shifts tend to cancel.

On a non-inverting transponder the downlink frequency is given by:

$$F_d = F_u - F_t$$

where F_t is the translation frequency. Signals in the downlink band have the same frequency order as those in the uplink band. This is shown in Figure 9.4(a).

On an inverting transponder the downlink frequency is given by:

$$F_d = F_t - F_u$$

and signals in the downlink band are transposed relative to those in the uplink band. This is shown in Figure 9.4(b).

Due to the motion of the satellite relative to the ground station there is a frequency shift of the received signal due to the Doppler effect. This shift is

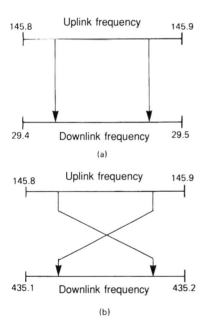

Figure 9.4 Relationship between uplink and downlink frequencies for a satellite transponder: (a) non-inverting transponder; (b) inverting transponder

proportional to the frequency of the signal, and for low-orbit satellites the typical worst case values for the amateur bands are:

Band (MHz)	Doppler shift
29	700 Hz
144	3 kHz
432	10 kHz
1260	29 kHz

The maximum shift depends upon orbit height, with the shift reducing as the height increases. For a non-inverting transponder the shifts for the uplink and downlink must be added together. For an inverting transponder the difference between the shifts for the uplink and downlink will apply. For an elliptical-orbit satellite (e.g. OSCAR 10) the Doppler shift at apogee is typically about an eighth of that for perigee or for a low-orbit satellite.

Amateur satellite operating modes

The operating mode of a satellite defines the frequency bands used for the uplink and the downlink.

Mode A	Uplink	145.850–145.950 MHz
	Downlink	29.400–29.500 MHz
Mode B	Uplink	435 MHz
	Downlink	145 MHz
Mode C	Low-power mode B	
Mode D	Telemetry only	
Mode J	Uplink	145.800–146.000 MHz
	Downlink	435.000–438.000 MHz
Mode JA	Uplink	145.900–146.000 MHz
	Downlink	435.800–435.900 MHz
Mode JD	Uplink	145.850–145.910 MHz
	Downlink	435.910 MHz

Digital transponder for packet radio.
Four channels spaced at 20 kHz on 145 MHz.

Single channel downlink on 435 MHz

Mode JL	Uplink	145 and 1215 MHz
	Downlink	435 MHz
Mode K	Uplink	21 MHz
	Downlink	29 MHz
Mode KA	Uplink	21 and 145 MHz
	Downlink	29 MHz
Mode KT	Uplink	21 MHz
	Downlink	29 and 145 MHz

Mode L	Uplink	1260.00–1270.00 MHz
	Downlink	435.00–438.00 MHz
Mode S	Uplink	435 MHz
	Downlink	2305 MHz
Mode T	Uplink	21 MHz
	Downlink	145 MHz

On the new OSCAR Phase 3D satellite which is expected to be launched in 1995 a different mode designation scheme is to be used. On this satellite the receivers and transmitters are linked by a switch matrix so that any receiver can be connected to any of the transmitters. Each band is given a code letter and the mode is specified by a two-letter group where the first letter defines the uplink band and the second letter the downlink band.

The letters for the various bands are:

C	5650.0–5670.0 MHz	Uplink
L	1260.0–1270.0 MHz	Uplink
S	2400.0–2450.0 MHz	
U	435.00–438.00 MHz	
V	144.00–146.00 MHz	
X	10 450–10 500 MHz	

For example, the old mode B (435 up/145 down) would become mode UV and the old mode L (1290 up/435 down) would become LU.

Amateur satellite band plans

On the OSCAR 13 satellite the 150-kHz-wide downlink band is divided into three segments for CW only, mixed CW/SSB and SSB only. This is shown in Figure 9.5. Beacons and telemetry are located at each end of the downlink band.

Russian satellites, which generally have a narrower transponder bandwidth, use a similar band plan with CW signals at the lower frequency end. It should be noted that on a reverse transponder the CW signals are transmitted in the upper part of the uplink passband.

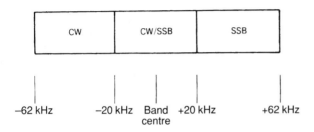

Figure 9.5 The general band plan arrangement used for OSCAR satellites

On the new Phase 3D OSCAR satellite it is expected that half of the transponder bandwidth will be allocated for packet radio and the other half for normal communications using CW or SSB.

Amateur satellite telemetry

Satellites normally carry a number of beacon transmitters which allow the satellite to be detected as it comes into range of the ground stations. In some cases the beacons in use will also indicate which transponders are operational.

All amateur satellites transmit telemetry signals which give the status of various parameters within the satellite. These signals are normally sent using Morse code on the satellite beacon frequencies. The telemetry consists of a string of numbers giving data for the various channels in sequence.

OSCAR Phase 2 satellites

These satellites carried communications transponders and telemetry transmitters. They were placed into a near-circular low orbit and provided relatively short periods of communications as they passed over a ground station. Examples were OSCAR 7 and OSCAR 8, which have now ceased operation.

OSCAR Phase 3 satellites

These satellites are launched into an elliptical orbit which permits relatively long periods of communications when the apogee of the satellite is on the same side of the earth as the ground station. Communications transponders, beacons and telemetry are provided on these satellites.

OSCAR 10 (AO10)

This was the first of the Phase 3 elliptical orbit satellites, and has transponders for modes B and L.

Mode B transponder (inverting)
Uplink	435.025–435.175 MHz
Downlink	145.975–145.825 MHz
Beacons	145.812, 145.900 MHz

Mode L transponder (inverting)
Uplink	1269.050–1269.850 MHz
Downlink	436.950–436.150 MHz
Beacons	436.040, 436.020 MHz

OSCAR 13 (AO13)
This was the second of the Phase 3 satellites to be placed in orbit. It carries mode B, mode J, mode L and mode S analogue transponders for normal CW and SSB communication.

Mode B transponder
Uplink	435.423–435.573 MHz
Downlink	145.975–145.825 MHz
Beacons	145.812, 145.975 MHz

Mode J transponder
Uplink	144.423–144.473 MHz
Downlink	435.990–435.940 MHz
Beacons	435.651 MHz

Mode L transponder
Uplink	1269.351–1269.731 MHz
Downlink	436.005–435.677 MHz
Beacons	435.651 MHz

Mode S transponder
Uplink	435.602–435.635 MHz
Downlink	2400.715–2400.749 MHz
Beacon	2400.664 MHz

OSCAR Phase 3D
In 1995 it is expected that a new OSCAR 3D satellite will be launched to replace OSCAR 13. This new satellite will have much more powerful transmitters and better antenna gain, so that its signals are expected to be some 10–20 dB better than those from OSCAR 13. The satellite also uses a switch matrix to connect the on-board receivers and transmitters so that any mode can be selected at will. Some of the transponders will have bandwidths of up to 500 kHz, compared with the 50-kHz bandwidth provided on OSCAR 13. It is expected that a large part of the activity on this satellite will use packet radio at 9600 baud or possibly higher rates on some transponders. The satellite is also expected to carry TV cameras for earth imaging and for star field images.

UOSAT satellites
These are produced by the university of Surrey and the UK AMSAT organization for scientific research experiments. UOSAT1 (OSCAR UO9) carried telemetry and beacon transmitters and an imager camera to give pictures of the earth's surface. This satellite was not intended for communications

purposes and has now ceased to operate. UOSAT 2 (OSCAR UO11) is similar to UOSAT 1 but carries different experiments.

UOSAT1 (UO9)
This was the first of the UOSATs, and has now ceased to operate.

HF beacons: 7050, 14 002 kHz CW
               21 002, 29 510 kHz CW
VHF beacon/telemetry: 145.825 MHz FM
UHF beacon/telemetry: 435.025 MHz FM
S Band beacon/telemetry: 2401.0 MHz FM
Microwave beacon: 10 470 MHz CW
Imager system: 256 × 256 pixels
Digital code with 4 bits per pixel
Uses VHF/UHF beacon Tx
Period: 95.3 min
Inclination: 97.48°

UOSAT 2 (UO11)
VHF beacon/telemetry: 145.825 MHz FM
UHF beacon/telemetry: 435.025 MHz FM
S band beacon/telemetry: 2401.5: MHz FM
Period: 98.56 min
Inclination: 98.25°

UOSAT (OSCAR UO22)
This UOSAT carries a CCD imager camera similar to that of UOSAT1. Telemetry and camera video share the same downlink using 9600-baud FSK data.

Downlink: 437.120 MHz

Japanese FUJI amateur satellites

Fuji 1 (OSCAR FO12)
This first Japanese satellite, Fuji 1 (OSCAR FO12), is designed to act as a communications satellite using mode JA. It also carries a mode JD packet radio repeater with a store and forward message handling facility.

Mode JA transponder
Uplink   145.900–145.933 MHz SSB
              145.933–145.967 MHz CW/SSB
              145.967–146.000 MHz CW

Downlink   435.900–435.967 MHz SSB
                 435.867–435.833 MHz CW/SSB
                 435.833–435.800 MHz CW

The Mode JA transponder is an inverting type.

Mode JD transponder
Uplink   145.850 MHz (Channel 1)
                 145.870 MHz (Channel 2)
                 145.890 MHz (Channel 3)
                 145.910 MHz (Channel 4)
Downlink   435.910 MHz
Satellite period: 115.8 min
Inclination: 50°
Height: 1498 km

Fuji 2 (OSCAR FO20)
This was the second Japanese Fuji-type satellite, and operates in mode JA and mode JD, similar to Fuji 1.

Mode JA transponder
Uplink   145.900–146.000 MHz
Downlink   435.900–435.800 MHz

Mode JD transponder
Uplink   145.850 MHz (Channel 1)
                 145.870 MHz (Channel 2)
                 145.890 MHz (Channel 3)
                 145.910 MHz (Channel 4)
Downlink   435.910 MHz
Data rate: 1200 baud PSK

Russian amateur satellites
Russian amateur satellites are given code names RS followed by a serial number. Early satellites used mode A transponders but the later satellites from RS10 onward use modes A, K and T. Early satellites from RS1 to RS9 have now ceased operation.

Each of these satellites carries a Robot system which provides automatic Morse contacts. The RS11 Robot transmits on one of its 29-MHz beacon frequencies using the format 'CQ de RS11 QSU 21300 AR' and listens on 21 300 kHz for calls which should have the former 'RS11 de (your call) AR'. The Robot responds by sending the callsign received and a QSO number. The Robots on the other RS satellites operate in a similar fashion. Telemetry signals are sent using Morse code.

RS10 and RS11 may also use two additional transponder modes. In mode KA the 21- and 145-MHz uplink signals are combined to give a single downlink on 29 MHz. For mode KT the uplink is on 21 MHz and the downlink is transmitted on both 29 MHz and 145 MHz at the same time.

RS10
Transponders are of the non-inverting type.

Mode A transponder
Uplink   145.860–145.900 MHz
Downlink   29.360–29.400 MHz

Mode K transponder
Uplink   21.160–21.200 MHz
Downlink   29.360–29.400 MHz

Mode T transponder
Uplink   21.160–21.200 MHz
Downlink   145.860–145.900 MHz

Robot transponder
Uplink   21.120–145.820 MHz
Downlink   29.357 or 29.403 MHz

Beacon/telemetry
29.357, 29.403 MHz
145.857, 145.903 MHz

RS11
Transponders are of the non-inverting type.

Mode A transponder
Uplink   145.910–145.950 MHz
Downlink   29.410–29.450 MHz

Mode K transponder
Uplink   21.210–21.250 MHz
Downlink   29.410–29.450 MHz

Mode T transponder
Uplink   21.210–21.450 MHz
Downlink   145.910–145.950 MHz

Robot transponder
Uplink   21.130–145.830 MHz
Downlink   29.407 or 29.453 MHz

Beacon/telemetry
29.407, 29.453 MHz
145.907, 145.953 MHz
RS10 and RS11 share the same space vehicle with an orbital period of 105.2 min and inclination of 82.93° and an increment of 26.26° on each orbit.

RS12 and RS13
Like RS10 and RS11, these two satellites share space on one of the Russian Cosmos navigation satellites. Their characteristics are similar to those of RS10 and RS11.

RS14 (OSCAR 21)
This Russian satellite, unlike earlier models, operates with two inverting transponders for mode B and a packet BBS system.

Mode B transponder
Uplink 1   435.102–435.022 MHz
Uplink 2   435.123–435.043 MHz
Downlink 1   145.852–146.932 MHz
Downlink 2   145.866–146.945 MHz
Beacons   145.822, 145.948 MHz
Telemetry   145.952, 145.838 MHz

Packet BBS
Uplink   435.016 MHz
                  435.041 MHz
                  435.155 MHz
                  435.193 MHz
Data rate   1200, 2400 baud PSK
                  4800, 9600 baud NRZI
Downlink   145.983 MHz
Data rate   1200, 2400, 4800, 9600 baud PSK

Microsat amateur satellites
Seven microsats (OSCAR 14–19) were launched in 1990. These small satellites provide various educational experiments or packet radio facilities.

UOSAT (OSCAR UO14)
This is a packet store and forward satellite with one channel.

Uplink	145.975 MHz
Downlink	435.070 MHz
Data rate	1200 baud AFSK FM or 9600 baud FSK

UOSAT (OSCAR UO15)
This is a packet microsat similar to UO14

PACSAT (OSCAR AO16)
This microsat provides a basic packet mailbox and BBS which operates at 1200 baud and uses PSK as a modulation method.

Uplink	145.900 MHz
	145.920 MHz
	145.940 MHz
	145.960 MHz
Downlink	437.025 MHz
	437.050 MHz
Data rate	1200 baud PSK

DOVE (OSCAR 17)
This microsat provides a basic educational facility with a simple digitized speech downlink on 145.825 MHz and a beacon on 2401.22 MHz.

WEBERSAT (OSCAR WO18)
This microsat contains a TV camera and transmits the pictures as digital data in packet format. There is also an ATV repeater which can store and retransmit an NTSC FS TV picture.

TV data downlink	437.102 MHz
Data rate	1200 baud PSK
TV repeater	1265 MHz

LUSAT (OSCAR LO19)
This Argentinian microsat carries a packet mailbox or BBS similar to that on PACSAT AO16. It has four 145-MHz uplink channels and a downlink at 437 MHz. Packet data are at 1200 baud and use PSK.

Uplink	145.840 MHz
	145.860 MHz
	145.880 MHz
	145.900 MHz
Downlink	437.125 MHz
	437.154 MHz
Data rate	1200 baud PSK

ARSENE (OSCAR AO24)

This is a French amateur satellite carrying a packet repeater system with three uplink frequencies at 435 MHz and downlinks on 145 and 2445 MHz. The packet system uses AX25 protocol at 1200 baud with FSK modulation.

Uplink	435.050 MHz
	435.100 MHz
	435.150 MHz
Downlink	(Mode B) 145.975 MHz
	(Mode S) 2445.54 MHz
Data rate	1200 baud FSK

During 1993 seven microsats were launched by an Ariane rocket. Of these, four carry amateur radio facilities.

ITAMSAT (OSCAR AO26)

This is an Italian amateur microsat designed for store and forward packet operation. In effect it is a BBS operating at 1200 baud, similar to those on PACSAT and LUSAT.

Uplink	435.867 MHz primary
	435.822 MHz secondary

Both uplink frequencies use 1200 baud PSK. The secondary frequency may also operate at 1200 baud AFSK (FM) or 9600 baud FM.

Downlink	145.875 MHz
	145.900 MHz
	145.925 MHz
	145.950 MHz

These downlink frequencies operate at 1200 baud normally but the lower frequency pair may also operate at 4800 baud and the upper pair may use 9600 baud.

POSAT (OSCAR 28)

This is a commercial Portuguese microsat built by the University of Surrey to handle earth images and radiation experiments but it also carries amateur transmitters. It will operate store and forward packet similar to a BBS.

Uplink	145.925 MHz
	145.975 MHz
Downlink	435.250 MHz
	435.275 MHz
Data rate	9600 baud
Commercial downlink	429.950 MHz
	429.450 MHz

EYESAT-A (OSCAR 27)

This is a commercial microsat, produced by AMRAD in Washington, which contains an amateur radio packet module.

Uplink	145.850 MHz
Downlink	436.800 MHz
Data rate	1200 or 9600 baud

KITSAT-B (OSCAR 25)

This Korean microsat carries a CCD imager for pictures of the earth and a mailbox or BBS packet system. Other experiments carried include an infrared sensor and a digital signal processor.

Uplink	145.870 MHz
	145.980 MHz
Downlink	435.175 MHz
	436.500 MHz
Data rate	9600 baud

Polar orbiter weather satellites

One form of satellite signal reception which can be of interest is the reception of FAX pictures from the various weather satellites that are in orbit. Many of these transmissions are made in the 137-MHz band and are easily received with a modified VHF converter. An international system known as WEFAX also provides satellite picture signals and operates at a frequency of 1691 MHz. Images from the spacecraft are initially transmitted to ground using a special high-resolution mode and then converted at the ground station to the standard FAX mode and sent back to the satellite, from which the picture is rebroadcast for reception by simpler ground stations.

Weather satellites fall into two major groups. The first are the low-altitude satellites which tend to have near-polar orbits and transmit downlink signals in the 137-MHz band. The US satellites of this type are called the NOAA series, whilst the Russian satellites of this type are the METEOR series. There are some differences in the formats of the FAX signals from the US and Russian satellites, but they both use 120 lines per minute and the video signal uses amplitude modulation of an AF subcarrier which is then used to frequency modulate the 137-MHz carrier signal.

The NOAA satellites normally transmit a pair of pictures side by side, with one giving the visual image and the second an infrared image. Russian satellites usually transmit one picture image and this is generally a visual image.

The received FAX picture can be reproduced by using a conventional FAX machine but for amateur stations it is more common to use some form of computer display system. This might use one of the popular home computers to provide a high-resolution graphics picture with, perhaps 8 or 16 grey levels and a resolution of, say, 256 by 256 pixels. More advanced home computers may permit higher resolution or more grey levels to give improved pictures. A permanent copy of the received picture may then be obtained by dumping the graphics display to a dot matrix, inkjet or laser printer.

It is also possible to actually store the picture using even higher resolution of perhaps 1024 by 1024 pixels and then display part of the picture at a time with high resolution or the whole picture with reduced resolution. The main constraint here is the amount of computer memory that is available for storing the picture information. An alternative approach is to use a dedicated display system which basically consists of a large digital memory to act as a frame store and a digital graphics display which will generally use some form of TV monitor.

Apart from the picture transmissions, the weather satellites may also transmit weather data using RTTY signals, where the data are in the form of groups of numbers following a similar format to that used by ground-based weather transmissions.

Currently, NOAA9, NOAA10, NOAA11 and NOAA12 are operational. The Russian satellites are the Meteor 3 series METEOR3-3, METEOR3-4 and METEOR3-5.

Frequencies (MHz) used for picture transmission from these low-level satellites are as follows:

NOAA9	137.620
NOAA10	137.500
NOAA11	137.620
NOAA12	137.500
METEOR3-3	137.850
METEOR3-4	137.300
METEOR3-5	137.300

Orbit predictions and transmission schedule for the NOAA satellites are broadcast by RTTY from Bracknell on 4489 kHz daily between 1930 and 2030 GMT.

Geostationary weather satellites

The second type of weather satellite operates in a geostationary orbit over the equator. The satellites transmit FAX pictures for the WEFAX service. The speed used is 240 lines per minute using the APT transmission format. Six of these satellites are eventually planned to be in orbit to give world-wide coverage, and have the following orbit positions:

METEOSAT 6   0° Europe
METEOSAT3   75° W America
GOES East   75° W USA
GOES Central   105° W USA
GOES West   130° W USA
GOMS   70° E USSR
GMS   140° E Japan

The transmission frequency is normally 1691.000 MHz but the METEOSAT frequency is 1694.500 MHz.

Spacecraft communications

Another type of signal that may be received from space is the communications downlink between a spacecraft such as the US space shuttle or the Russian Soyuz craft or the MIR space station and their ground control stations. Russian spacecraft use frequencies around 142 MHz. US shuttle transmissions may sometimes be found in the 250–300-MHz band, although most shuttle communications are made via geostationary relay satellites using the C band.

From time to time, MIR and various shuttle missions carry amateur radio stations, which usually operate in the 145-MHz band.

Television satellites

Some enthusiasts search the skies for signals from the many TV and communications satellites that are visible from Europe. Although some satellites, such as Astra, broadcast TV entertainment channels, the other satellites carry news feeds, programme feeds and video conferences as well. These transmissions do not run to any regular schedule and are found by searching through the available channels on each satellite.

Televison signals are transmitted via satellites operating in the 3.5–4.5-GHz (C) band and 10.75–12.75-GHz (Ku) band microwave frequencies. The C band communications satellites are used for intercontinental links between broadcasting networks and for feeds to cable networks. Signal levels from these satellites are low and require large dish antennas at the receiving station. Typically these are some 2–7 m in diameter, although some signals may be receivable on smaller dishes.

Television signals transmitted by the Ku band satellites are generally much stronger and most can be received by using dishes of 90 cm or 1.2 m diameter within their general service area. A motorized dish antenna which can be moved to point at any of the visible satellites is essential. The antenna position controller may be built into the receiver unit or may be in a separate unit. Signals from direct broadcast satellites and some other satellites, such as Astra, can be received on dishes as small as 60 cm in diameter.

The Ku band is effectively divided into three sub-bands as follows:

Fixed satellite service (FSS) band	10.75–11.70 GHz
Direct broadcast (BSS) band	11.70–12.50 GHz
Fixed satellite service (FSS) band	12.50–12.75 GHz (Telecom)

Each communications satellite carries up to 24 separate transponders, each having a bandwidth of some 36 MHz. In many cases a transponder is occupied by a single TV channel but on some satellites two TV signals may be carried using one transponder. Transponders which are not being used for TV carry other services such as intercontinental telephone channels, and various other radio channels.

On C band satellites the uplink frequencies are in the range 5945–6405 MHz and downlink frequencies run from 3720 to 4380 MHz. The centre frequencies of the 24 channels in this band are spaced 20 MHz apart. Because each channel is 36 MHz wide, the adjacent channels overlap in frequency, so their signals are polarized in different directions (e.g. vertically and horizontally) to reduce interference. Channels with the same polarization are spaced 40 MHz apart.

On Ku band the downlink for communications satellites is in the range 10.95–11.70 GHz. A recent extension to this band has its downlink in the range 10.70–10.95 GHz. For direct broadcast satellites the downlink is in the range 11.70–12.50 GHz.

At present, amateur space satellites do not cater for TV transmissions but it is possible that future satellite projects may include a transponder for FSTV signals.

For satellite transmission the TV signal is frequency modulated onto its carrier, and several sound subcarriers may be included to provide different language sound channels. Sound subcarrier frequencies on the channels vary from 5 to 8 MHz and are generally FM. Most video signals currently use one of the normal broadcast standards with colour by NTSC, PAL or SECAM methods. Some European channels use DMAC or D2MAC and can provide widescreen 16 : 9 format pictures. In the next few years digital transmissions using a version of the MPEG2 data compression system will gradually take over for satellite broadcast transmissions.

Communications-type satellites are often used to provide TV services for cable networks or paying subscribers, and some form of scrambling may be used to prevent unauthorized viewing. On the video signal this usually takes the form of altering the sync. pulse so that a conventional TV set cannot lock the picture correctly, whilst for sound some form of analogue scrambling technique may be used. Some channels broadcast the sound as a digital signal within the sync. pulse period, and this digital signal may be enciphered before transmission so that only receivers with an appropriate decoder unit can

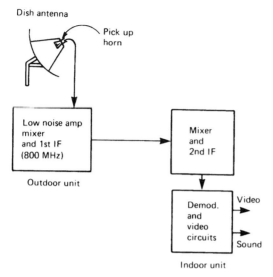

Figure 9.6 Basic system for reception of satellite television signals

decipher the sound. Many direct broadcast channels are scrambled by using the Videocrypt system, which cuts each scan line into segments and then shuffles them. At the receiving end a Videocrypt decoder is used which requires a smart card to be inserted before it will decode the scrambled signals. On DMAC systems the Eurocrypt encryption system is used, which operates in a similar fashion.

The receiving station for satellite TV consists of a dish antenna which focuses the signal onto a feed horn mounted in front of the dish. Signals from the feed horn are fed to a low-noise amplifier and converter, which produces an IF in the region of 950–2000 MHz. This signal is fed via cable to an indoor receiver which contains the channel tuning, an IF amplifier with a bandwidth of the order 25 MHz and an FM demodulator to produce the video output and sound subcarriers. In most systems the video and sound are remodulated onto a UHF carrier such as channel 36 and then fed to the aerial input of a standard TV receiver.

In the multiplexed analogue components (MAC) system the luminance and chrominance signals are transmitted separately by using a time multiplexing system. The luminance signal is compressed to fit into two-thirds of the line scan period and the chrominance signal is sent during the remaining one-third of the line period. Synchronization and sound make use of digital signals in the line blanking period at the start of each line scan. The compression and expansion of the analogue components of the video signal are achieved by using digital techniques, and special integrated circuits are available to provide the signal processing required in the receiver.

For MAC-type transmissions and scrambled signals a decoder unit is included after the main IF detector to extract the video and sound signals and convert them to a form suitable for a normal TV receiver.

European Ku band TV satellites

Location	Name	Signals
63°E	Intelsat 602	TV and feeds
60°E	Intelsat 604	TV and feeds
48°E	Eutelsat 1 F1	TV and communications
42°E	Turksat	Turkish TV (1994)
28.5°E	Kopernicus DFS2	German TV links
25.5°E	Eutelsat 1 F4	News feeds
23.5°E	Kopernicus DFS3	German TV
21.5°E	Eutelsat 1 F5	News feeds
19.2°E	Astra 1A	16 TV channels
19.2°E	Astra 1B	16 TV channels
19.2°E	Astra 1C	18 TV channels
19.2°E	Astra 1D	Digital TV (1994)
19.2°E	Astra 1E	Digital TV (1995)
16.0°E	Eutelsat 2 F3	16 TV channels
13.0°E	Eutelsat 2 F1	16 TV channels
13.0°E	Eutelsat 2 F6	16 TV channels (1994)
10.0°E	Eutelsat 2 F2	16 TV channels
7.0°E	Eutelsat 2 F4	Eurovision TV feeds
5.0°E	Tele X	Scandinavian TV
5.0°E	Sirius	Scandinavian TV
3.0°E	Telecom 1C	French news feeds
1.0°W	Intelsat 512	Scandinavian TV
1.0°W	Intelsat 702	Scandinavian TV (1994)
1.0°W	Thor	Scandinavian DBS TV
5.0°W	Telecom 2B	French TV and feeds
8.0°W	Telecom 2A	French TV
11.0°W	Statsionar 11	News feeds from Russia
14.0°W	Statsionar 12	News feeds from Russia
18.0°W	Intelsat 515	Some TV and news feeds
19.0°W	TV Sat 2	German DBS satellite
19.0°W	TDF 1	French DBS satellite
21.5°W	Intelsat K	News feeds
27.5°W	Intelsat 601	TV and news feeds
31.0°W	Hispasat 1A	Spanish TV
31.0°W	Hispasat 1B	Spanish TV
43.0°W	Panamsat PAS3	TV and feeds (1994)
45.0°W	Panamsat PAS1	Mexican TV and news feeds
50.0°W	Intelsat 506	NBC programme feeds

European DBS television channels
The European DBS band is divided into 40 channels, each 36 MHz wide and spaced at 19-MHz intervals. Adjacent channels overlap in frequency but interference is avoided by arranging that adjacent transponders use a different antenna polarization.

(a)

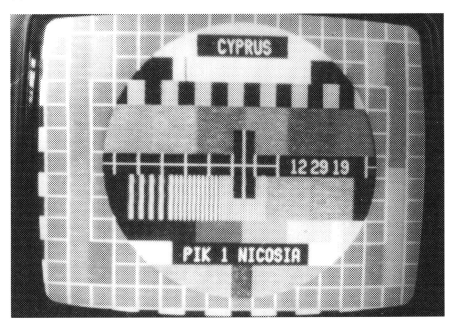

(b)

Figure 9.7 Examples of satellite television signals

(c)

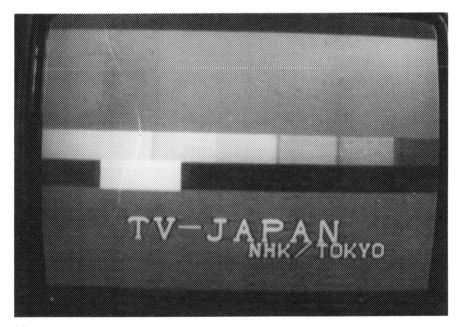

(d)

Channel no.	Frequency (GHz)
01	11.727
02	11.746
03	11.765
04	11.785
05	11.804
06	11.823
07	11.842
08	11.861
09	11.880
10	11.900
11	11.919
12	11.938
13	11.957
14	11.976
15	11.995
16	12.015
17	12.034
18	12.053
19	12.072
20	12.091
21	12.111
22	12.130
23	12.149
24	12.168
25	12.187
26	12.206
27	12.226
28	12.245
29	12.264
30	12.283
31	12.302
32	12.322
33	12.341
34	12.360
35	12.379
36	12.398
37	12.417
38	12.437
39	12.456
40	12.475

European DBS television channel allocations

Country	Orbit slot	Channels
Belgium	19°W	21, 25, 29, 33, 37
UK	31°W	4, 8, 12, 16, 20
Denmark	5°E	12, 15, 20, 27, 35
Finland	5°E	2, 6, 10, 22, 26
France	19°W	1, 5, 9, 13, 17
Germany	19°W	2, 6, 10, 14, 18
Ireland	31°W	2, 6, 10, 14, 18
Luxembourg	19°W	3, 7, 11, 15, 19

Netherlands	19°W	23, 27, 31, 35, 29
Norway	5°E	14, 18, 28, 32, 38
Portugal	31°W	3, 7, 11, 15, 19
Spain	31°W	23, 27, 31, 35, 39
Sweden	5°E	4, 8, 30, 34, 40

News feed and video link channels

Eutelsat 1F4 25.5°E
11 134 MHz. H Racing and news feeds
11 171 MHz. H Racing and news feeds

Eutelsat 2F3 16°E
10 987 MHz. H Eurostep education programmes
11 165 MHz. H Business TV and news feeds
12 540 MHz. V News feeds
12 584 MHz. V News feeds

Eutelsat 2F1 13°E
11 616 MHz. V Business TV and news feeds
11 596 MHz. H Sports feeds
11 678 MHz. H Sports feeds
12 520 MHz. H Reuters news feeds
12 564 MHz. H World-wide TV news
12 584 MHz. V News feeds and sport

Telecom 1C 3°E
12 606 MHz. V Conference TV and news feeds
12 648 MHz. V News and sports feeds

Telecom 2B 5°W
12 548 MHz. H Sports and news feeds
12 586 MHz. H Sports and news feeds

Stationar 14°W
12.525 MHz. R Reuters news, Moscow

Intelsat K 21.5°W
11.471 MHz. H German news feeds
11 499 MHz. H Reuters news feeds
11 530 MHz. V Reuters news feeds
11 532 MHz. H News feeds from US
11 558 MHz. V Reuters and CNN news feeds

Intelsat 601 27.5°W
11 015 MHz. H News feeds from US
11 475 MHz. V EBU news feeds from US

Panamsat PAS1 45°W
11 596 MHz. H NHK Japan news feeds
11 639 MHz. H CBS news feeds from US
11 675 MHz. H NHK programme feeds

Intelsat 506 50°W
11 638 MHz. V NBC programme feeds

Amateur moonbounce (EME) contacts

Some amateur stations use the surface of the moon as a passive satellite reflector to make long-distance contacts on the VHF and UHF bands. This requires highly directional dish-type antennas and very low noise receivers.

Frequencies used for earth–moon–earth (EME) operation (MHz) are:

144.000–144.025
432.000–432.025
1296.000–1296.025

10

The broadcast bands

One aspect of the listener's hobby is that of receiving broadcast stations on the short-wave broadcast bands. Stations can be received from many countries in the world and one aim of the listener might be to log as many different countries as possible. The more interesting stations are those which are intended for more local reception and are often difficult to receive.

Although most of the major international broadcasters usually have some of their programmes in English, the Dx stations will often broadcast in local languages and sometimes they can be identified only by their frequency and perhaps by the type of programme being transmitted unless the listener understands the language being used. International broadcasters such as the BBC, Voice of America and Radio Moscow usually transmit on several different bands simultaneously and are easy to find.

International short-wave broadcasting bands
For international broadcasting the following bands are generally used:

75-m band	3950–4000 kHz
49-m band	5959–6200 kHz
41-m band	7100–7300 kHz
31-m band	9500–10 000 kHz
25-m band	11 650–12 050 kHz
22-m band	13 600–13 800 kHz
19-m band	15 100–15 600 kHz
16-m band	17 550–17 900 kHz
13-m band	21 450–21 850 kHz
11-m band	25 600–26 100 kHz

The 75-m band is a regional band used in Europe and Asia.

The 11-, 13-, 16- and 22-m bands are mainly used by the major international broadcasters. The 19-, 25-, 31-, 41- and 49-m bands contain many lower powered and local stations which will be of more interest to Dxers.

Although many stations transmit in English at various times, most programmes are likely to be in other languages. In some cases it may only be possible to identify a station from its frequency and the language or type of programme. Most languages can be identified by their characteristic sound. Thus although French, Spanish, Italian and Portuguese are all based on Latin, they can readily be identified even if you do not understand what is being said. Having identified the language, look up the possible stations on the frequency and see which one is likely to be broadcasting in that language at the time it is received.

Tropical broadcast bands

Some of the lower frequency short-wave bands allocated for broadcasting are used only by stations in the tropical region between 23N and 23S latitudes and are for local broadcasting in Africa, Asia and parts of Central and South America. These are generally referred to as the tropical bands, and can provide some interesting Dx signals when conditions are good. Various utility stations also share these bands, so interference with the broadcast signals is often a problem in Europe and tropical storms also cause severe static interference on these bands.

The tropical bands are:

120-m band	2300–2498 kHz
90-m band	3200–3400 kHz
60-m band	4750–5060 kHz

Some of the stations which may be heard in Europe are:

ABC, Alice Springs, Australia	2310 kHz
ABC, Katherine, Australia	2494 kHz
Radio Namibia	3270 kHz
Freetown, Sierra Leone	3315 kHz
Radio Nigeria	3326 kHz
Radio Botswana	3356 kHz
GBC Ghana	3366 kHz
Radio Brazzaville, Congo	4765 kHz
Radio Nigeria, Kaduna, Nigeria	4770 kHz
Radio Lesotho	4800 kHz
Radio Yerevan, Armenia	4810 kHz
RTV, Burkina Faso	4815 kHz
Radio Kharkov, Ukraine	4825 kHz
Radio Bamako, Mali	4835 kHz
Radio Nouakchott, Mauritania	4845 kHz
Radio Cabocla, Manaus, Brazil	4845 kHz
Radio Yaounde, Cameroon	4850 kHz

Radio Cotonu, Benin	4870 kHz
Radio Beijing, China	4883 kHz
Radio Conakry, Guinea	4892 kHz
Radio Zambia	4910 kHz
Radio Accra, Ghana	4915 kHz
Voice of Kenya	4935 kHz
Radio Kiev, Ukraine	4940 kHz
Radio Baku, Azerbaijan	4960 kHz
RSA, South Africa	4965 kHz
Radio Ecos del Torbes, Venezuela	4985 kHz
Radio Nigeria, Lagos, Nigeria	4990 kHz
Radio Alma Ata, Kazakhstan	5035 kHz
Radio Lome, Togo	5047 kHz
Radio Cayenne, French Guiana	5055 kHz

In Europe, stations from Africa will usually be heard during the early part of the evening. Stations from South and Central America are generally audible from around midnight until dawn. Asian stations may be heard during the afternoon when conditions are favourable. Reception on these bands is usually best during winter and improves during the sunspot minimum period.

Medium-wave broadcast Dxing

It is possible to hear broadcast stations from outside Europe on the medium-wave bands. Stations from North and South America may be heard during the winter months, although reception is becoming increasingly difficult as more of the powerful European transmissions now operate for 24 hours a day. On those channels where the more local European stations close down for a few hours during the night, it is possible to hear the weaker Dx stations.

Signals from the east coast of the USA and Canada become audible from around midnight in northern Europe. Reception is not easy because there is often deep fading and there may be two or more stations coming in on the same channel. Another problem is that there will often be crackles from static interference. Problems may be encountered with splatter from strong local stations on nearby frequencies but this can sometimes be dealt with by using a narrow-band receiver filter to reduce the bandwidth to perhaps 3 kHz instead of the more usual 6–10 kHz on an AM receiver.

Most keen medium-wave Dx listeners use some form of tuned loop antenna for reception rather than a long wire antenna. The loop has directional properties and provides a deep null in the direction along the plane of the loop with maximum pickup at right angles to the plane of the loop. A typical loop might be made, say, 3 feet square, and consists of about 10 turns of wire wound onto a wooden or plastic frame. The turns should be spaced about 0.5 cm apart to allow more turns to be used without having too much inductance. The loop is then tuned by a 500-pF variable capacitor and is peaked for maximum signal on the frequency being used. The loop is mounted vertically, and by rotating the loop it is possible to greatly reduce the strength of local European signals to allow the weaker Dx signals to be heard.

Coupling the loop to the receiver can be done by having a single-turn coupling winding mounted on the frame alongside the main tuned winding. This coupling loop is then fed to the antenna and earth terminals of the receiver. A better system is to use a balanced amplifier to provide the coupling. One problem with the loop is that the directional effect is produced when the loop works from the magnetic field of the incoming signal. A better method is to feed the signals from the main loop to the inputs of an operational amplifier and use the low-impedance output of the amplifier to drive the receiver. This allows any pickup of the electric field on the loop to be cancelled out and improves the depth of the null in the directional pattern of the loop. The amplifier also provides a high-impedance load across the

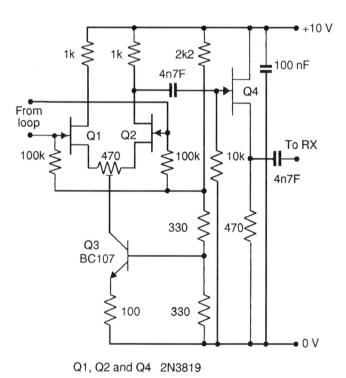

Q1, Q2 and Q4 2N3819

Figure 10.1 Circuit for a differential amplifier for use with a medium-wave loop antenna

loop which will give increased Q and better selectivity, which helps to reduce problems with splatter from strong European signals.

Some stations which may be audible under good conditions are:

VOCM, St Johns, Newfoundland	590 kHz
CKCM, Grand Falls, Newfoundland	620 kHz
CBF, Montreal, Quebec	690 kHz
Caribbean Beacon, Anguilla	690 kHz
WABC, New York	770 kHz
WHDH, Boston, Mass	850 kHz
CJCH, Halifax, Nova Scotia	920 kHz
CFBC, St John, New Brunswick	930 kHz
CBM, Montreal, Quebec	940 kHz
CHER, Sydney, Nova Scotia	950 kHz
WINS, New York	1010 kHz
CHUM, Toronto, Ontario	1050 kHz
WEVD, New York	1050 kHz
WBAL, Baltimore Md	1090 kHz
WNEW, New York	1130 kHz
WHAM, Rochester, NY	1180 kHz
WTOP, Washington DC	1500 kHz
WWKB, Buffalo, NY	1520 kHz

As an alternative to chasing medium-wave stations outside Europe, some listeners look for the low-powered local stations from inside Europe. With the arrival of commercial radio in the UK there are many local stations which can be heard well outside their local coverage area when conditions are favourable. Stations from Africa and the Middle East may also be found on medium wave under favourable conditions.

A directional loop antenna will be useful for this activity, since it will enable some of the stronger local signals to be reduced in strength to allow reception of more distant weak stations.

VHF broadcast band Dx

Some listeners use the VHF FM broadcast band as a hunting ground for Dx. In between the local broadcast stations it will usually be possible to hear some of the more distant stations, and when conditions are favourable reception of stations several hundred miles away becomes possible. The main FM broadcast band is 88.5–108 MHz, but some eastern European countries operate stations in the band 65–74 MHz.

For best results a beam antenna with a rotator system is desirable. Rotation may be controlled either by a motor at the top of the mast and a suitable controller and direction indicator or by some form of manually operated scheme.

Propagation conditions in this VHF band tend to be similar to those on the 2-m amateur band, except that distances over which reception is possible are several times greater due to the higher power of the broadcast stations together with their high antenna locations and the propagation characteristics of the lower frequency band.

Television Dx

Another aspect of broadcast Dx listening is that of chasing distant TV stations on VHF bands 1 and 3. For this activity a multi-standard TV receiver capable of handling PAL-G, PAL-I, SECAM and perhaps NTSC is desirable. A rotatable beam antenna is also a must for this activity and should be mounted as high as is practicable.

Long-distance TV reception often relies on tropospheric ducting, so it is useful to watch the weather charts for high-pressure conditions and the possibility of temperature inversion conditions. A scanner receiver set to sweep through the interesting channels can give an early indication of a possible opening when the buzz of the video signals will be heard coming up out of the noise or the sound channel may be detected in some cases.

11

Radio wave propagation

When a radio antenna is fed with an RF signal it generates magnetic and electric fields which then propagate away from the antenna in all directions. The way in which the radio waves propagate from the antenna varies depending upon the location of the antenna relative to ground and may be modified by reflections from the ionosphere or ducting effects within the troposphere. The study of radio propagation is a complex subject but here we shall look at the main features of radio propagation that will be of interest to the radio amateur and listener.

Free space propagation

If the antenna of a transmitter were located in free space and radiated equally in all directions, the signal strength at some distance r from the transmitter would be given by:

$$E = \frac{(30 \times P)}{D \times D} \text{ volt/metre}$$

where P is the transmitter power in watts and D is the distance in metres between transmitting and receiving antennas.

Here it will be seen that the signal strength falls off in inverse proportion to the square of the distance. This ideal situation would apply in the case of communications between two space vehicles.

Ground wave propagation

For most applications the transmitting and receiving antennas are located near the ground and the transmitted signal consists of a 'ground wave' travelling over the ground surface and a 'sky wave' which is radiated into the atmosphere.

Over a perfect conducting surface the ground wave falls off inversely with the square of the distance. In practice, if the ground is not a good conductor this causes further attenuation of the signal. At frequencies up to 200 kHz the attenuation is relatively small and propagation over thousands of kilometres is possible. At higher frequencies the ground wave attenuation increases rapidly with frequency and propagation range falls. At around 14 MHz the ground wave will extend for perhaps 20 km for an average amateur radio station.

Ground wave propagation is used during daylight for LF and MF band broadcasting, giving good signals up to a distance of perhaps 100 km. Ground wave radiation from a horizontal antenna is small, so MF broadcast stations generally use a vertical antenna. On the HF bands, ground wave propagation is limited to a few miles and is useful only for local contacts.

The ionosphere

For HF propagation over long distances the sky wave is used, because it can be reflected by the ionized layers in the upper regions of the earth's atmosphere. This ionized region of the atmosphere is known as the ionosphere and contains three major ionized layers which are referred to as the D, E and F layers (Figure 11.1). The effect of the E and F layers is to bend the path of the wave and to allow it to be reflected back to earth, thus giving communications paths which are much longer than those provided by the ground wave.

The D layer

The D layer is the lowest ionized layer and exists at a height of 70–90 km. Because it is relatively low in the atmosphere, the density of atoms in the D layer is still quite high. When a radio wave collides with an ionized atom it

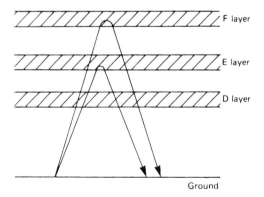

Figure 11.1 The D, E and F ionized layers in the earth's ionosphere

gives up some of its energy and therefore the field strength is attenuated. This attenuation effect in the D layer is greatest for the lower frequency signals and becomes less as the frequency increases. The D layer does not reflect radio signals but merely attenuates them as they pass through.

The ionization in the D layer is produced by radiation from the sun which excites the atoms within the layer. As a result the D-layer ionization exists only during daylight and is most intense at midday.

During the day the D layer absorbs almost all of the LF and MF signals, although a small part of these MF signals which enter the layer vertically will pass through and be reflected by the E layer. Thus it is possible to get reflected MF signals during daylight but the range is limited to little more than that of the ground wave. The HF, VHF and UHF signals pass through the D layer with some attenuation during daylight.

The E layer
Above the D layer at a height of 100–120 km is the E layer. This layer reflects signals in the MF and lower HF bands to provide propagation up to perhaps 150 km during daylight. Like that of the D layer, the ionization of the E layer is strongest at midday, but after dark the intensity of the E layer falls off more slowly and may persist through the night. At night this layer can provide long-distance propagation on the MF bands.

The F layers
The upper ionized layer is known as the F layer. During daylight it splits into two layers, F1 at about 150 km and F2 at 300 km. After dark the F1 and F2 layers merge and the ionization level falls slowly. The F layer provides the main long-distance propagation path for the HF bands, with lower frequency bands becoming more effective at night as the attenuation caused by the D layer falls away.

The effectiveness of the F layer follows a cyclic change through the year, giving best results on bands above 10 MHz during summer and below 10 MHz during winter. This layer is also affected by the level of activity of the sun and follows an 11-year cycle which is related to the number of sunspots on the sun's surface. At sunspot maximum the HF bands up to 28 MHz may remain open for long-distance contacts throughout the day and night.

Sporadic E propagation
In addition to the main ionized layers there is also an irregular ionization effect which occurs at the height of the E layer. The effect is that a thin but intense ionized layer appears within the region of the E layer and this allows signals in the bands 28–100 MHz to be reflected, giving long-distance propagation at these frequencies. This effect is known as 'sporadic E' propagation and occurs mainly during the summer months. This propagation mode can provide good Dx results on the 28-MHz band.

Sunspot cycles

The general pattern of propagation on the HF bands is greatly influenced by the activity of the sun. Sunspots have been observed for many centuries and have been found to follow an 11-year cycle. When sunspot activity is at a maximum level the propagation on the higher frequency bands becomes extremely good, since the upper layers of the atmosphere become heavily ionized. At the sunspot minimum the bands above about 15 MHz tend to be dead for long-range operation, although occasionally there will be brief openings during peaks of sunspot activity.

The last sunspot maximum occurred in 1991 and it is expected that the minimum of the current cycle will occur in 1996, after which the cycle should again rise to give a new peak around the year 2002.

Critical frequency (f_c)

The critical frequency f_c is the highest frequency at which a wave sent vertically up to the ionosphere is reflected by the E or F layer. The delay time between the transmitted pulse and the reflected echo gives a measure of the effective height of the layer (Figure 11.2).

Maximum usable frequency (MUF)

When the space wave enters the ionosphere layer its path is bent as it travels through the layer. At the lower frequencies this bending action is quite rapid and the wave path is turned so that the wave is reflected back from the ionosphere and returns to earth. As the frequency is increased, the amount of bending decreases, and the wave penetrates deeper into the layer before

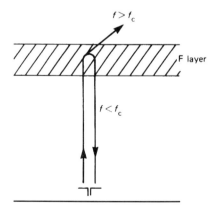

Figure 11.2 Ionospheric sounding technique used to find the critical frequency f_c and the height of the ionized layers

being reflected. At even higher frequencies the wave passes through the layer and goes on into space.

The highest frequency at which the wave is reflected from the ionosphere is known as the maximum usable frequency (MUF). To obtain a measure of the MUF it is normal to take ionospheric soundings to find the critical frequency f_c. The MUF is typically three times f_c for E-layer reflection and five times f_c for the F layer.

Commercial telecommunications circuits generally work at frequencies of about 0.85 of the MUF to ensure reliable communication, and this frequency is known as the optimum frequency.

Skip distance

The simplest form of ionospheric propagation involves a single reflection from the ionized layer. The antenna radiates waves over a range of vertical angles, so that the reflected wave returns to earth over a range of distances. At high vertical angles the signal may not be reflected back and this produces a 'skip zone' where there is no reflected signal and no signals from the transmitter can be received (Figure 11.3). The 'skip distance' to the nearest point at which the reflected signal can be heard depends upon the height of the reflecting layer and the angle at which the transmitted signal is launched. For maximum distance the angle of radiation should be as low as possible.

A single-hop transmission in which the signal is reflected only once from the E layer gives a maximum range of the order 2000 km. The higher F layer gives a single-hop range up to 4000 km. Greater distances are possible by multi-hop propagation, where the signal is reflected back and forth

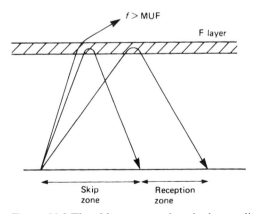

Figure 11.3 The skip zone produced when radio waves are reflected from the ionosphere

between the ionosphere and the ground. Each hop attenuates the signal but usable signals can be sent all the way around the earth via multi-hop propagation.

Amateur band propagation characteristics

160-m band (1.8 MHz)
This is the only MF band available to radio amateurs. During the day the band can give contacts up to about 100 km using either ground wave or reflection from the E layer. After dark, signals reflected by the E layer can give good Dx up to perhaps 8000 km, and under good conditions world-wide contacts are possible. The main problem on this band is the high level of atmospheric noise, which tends to be highest in summer. Propagation conditions tend to be best during winter months and around the period of the sunspot minimum, when D layer attenuation and noise are lower.

80-m band (3.5 MHz)
This band is less affected by the D layer and will permit contacts at distances up to around 4000 km during daytime. At night the band opens up for distances up to 20 000 km or more. Like 1.8 MHz, this band gives better results during winter, when atmospheric noise problems are reduced. One major problem on this band is that the skip distance is virtually zero, so that QRM from local stations makes working Dx signals very difficult.

40-m band (7 MHz)
This band can provide medium-range communications during daylight with contacts up to perhaps 1000 km. From dusk until dawn the band is capable of providing world-wide contacts, especially in the winter when atmospheric noise is low. At night the skip distance is typically 500 km, which reduces the amount of QRM from local stations.

30-m band (10 MHz)
This band can be useful throughout the day and night for long-distance working. During the day the skip distance is around 300 km, and contacts at distances up to 1500 km or more are possible. At night the band can open up for world-wide Dx as the skip distance widens to around 1000 km. The 10-MHz band is not greatly affected by the solar cycle, although in a solar minimum there will be some days when the band is dead because the MUF has dropped below 10 MHz.

20-m band (14 MHz)
This is the radio amateur's primary Dx band and is also extremely busy. During daylight this band can provide world-wide Dx contacts. During the

peak of the solar cycle the 14-MHz band will often remain open for Dx all through the night as well to provide a 24-hour Dx band. At the solar minimum the band only opens during daylight. Skip distances vary from around 750 km during the day to around 1500 km at night. The main problem on this band is QRM due to the high level of activity.

17-m band (18 MHz)
Like 14 MHz, this band can provide world-wide Dx during daylight hours but closes soon after sunset. At the solar minimum the band may be effectively closed on many days because the MUF is too low. A big advantage of this band is that when it is open the amount of activity tends to be less than on 14 MHz, so that QRM is much lower.

15-m band (21 MHz)
This band is primarily a daylight band, with signals falling off rapidly after dark. At solar maximum periods it can provide Dx and tends to be much quieter than 14 MHz in terms of QRM. During the solar minimum the band may be effectively closed for long periods for Dx activity.

12-m band (24 MHz)
This relatively new band for amateur radio has similar characteristics to the 21-MHz band, giving world-wide Dx contacts during daylight, particularly during the solar maximum period.

10-m band (28 MHz)
This band comes into its own during the peaks of the solar cycle, when it can provide world-wide Dx contacts even for low-power stations. This is primarily a daylight band. This band remains closed for Dx most of the time during sunspot minimum. This band can provide useful Dx contacts by sporadic E propagation during the summer months.

6-m band (50 MHz)
This VHF band can sometimes open up to give Dx contacts by reflection from the F layer during the period of solar maximum. Like 28 MHz, it is a daytime-only band. Long-range contacts can also be achieved in summer by sporadic E propagation.

Propagation beacons
In order to provide an indication of propagation conditions, a number of beacon stations have been set up on the higher frequency HF bands. These are generally fairly low power stations operating continuously or on a timeshared basis.

14-MHz amateur beacons

The 14-MHz beacons were set up during World Communications Year (1983) by the Northern California DX Foundation (NCDXF). All operate on 14 100 kHz using A1A on a timeshared basis. Each station operates in sequence for a period of 1 min. In the first 10 seconds the call sign is given in Morse. This is followed by 9-second dashes with power levels of 100 W, 10 W, 1 W and 100 mW in that order. The dashes are preceded by one, two, three or four dots to identify the power level. During the last 10 seconds the station again identifies with its callsign in Morse. The sequence of beacons is as follows:

Minute	Call	Location
00	4U1UN/B	New York, USA
01	W6WX/B	Palo Alto, California, USA
02	KH60/B	Oahu Island, Hawaii
03	JA1IGY	Mt Asama, Japan
04	4X6TU/B	Tel Aviv, Israel
05	OH2B	Helsinki, Finland
06	CT3B	Funchal, Madiera Islands
07	ZS6DN/B	Transvaal, South Africa
08	LU4AA/B	Buenos Aires, Argentina

The sequence starts on the hour with 4U1UN/B and cycles through in the order shown above with the cycle repeating every 10 min. The last two minutes are reserved for two additional beacons in South America. There are plans to extend the coverage by adding further beacons in the future when the cycle time may be increased to 15 or 20 min.

21-MHz amateur beacons

At the time of writing there are no beacons in this band but a series of time-multiplexed beacons similar to those on 14 MHz is planned. These are expected to operate on a frequency of 21 150 kHz.

28-MHz amateur beacons

In the band 28 200–28 300 kHz there are a number of fixed-frequency beacons operated by the various amateur organizations around the world. The operation of beacons in this band is currently under review and it is expected that in the future a number of timeshared beacon networks similar to those on 14 MHz will be set up and the fixed-frequency beacon system will be rationalized to provide a coordinated world-wide beacon system. Most of the currently operational 28-MHz beacons are low powered but they can provide a useful guide to the propagation conditions on this band. There are about 80 beacons in the 28-MHz band but the number varies as new beacons start up and others close down. A selection of the many 28-MHz beacons is given below as a guide.

Call	Frequency (MHz)	Location
A9XC	28.245	Bahrain
GB3SX	28.215	England
HG2BHA	28.2225	Hungary
LA5TEN	28.2375	Oslo, Norway
VE2TEN	28.2175	Quebec
VE3TEN	28.275	Ottawa
VE7TEN	28.2525	Vancouver
VK2WI	28.2625	Sydney
VK5WI	28.260	Adelaide
VP9BA	28.235	Bermuda
VS6HK	28.290	Hong Kong
VU2BCN	28.295	India
ZL2MHF	28.230	New Zealand
ZS1CTB	28.2425	Cape Town, South Africa
YV5AYV	28.280	Caracas, Venezuela
3B8MS	28.210	Mauritius

ITU HF beacons

As part of a study of HF propagation the ITU (International Telecoms Union) has set up a number of beacon transmitters in the HF bands. The beacons are multiplexed like the amateur beacons on 14 MHz but in this case they switch frequency in a time sequence. The frequencies used are 5470, 7870, 10 407, 14 405 and 20 945 kHz. Each station steps through the frequencies in ascending frequency order, stepping to a new frequency every 4 min. The cycle is then repeated every 20 min. Each station starts at a different point in the frequency sequence on the hour.

The stations currently operating are:

VK4IPS, Brisbane, Australia
LN2A, Svejo, Norway

The frequency sequences are as follows:

Min	VK4IPS	LN2A
00	5 470	14 405
04	7 870	20 945
08	10 407	5 470
12	14 405	7 870
16	20 945	10 407

This sequence repeats from 20 to 56 min and 40 to 56 min past the hour. Another three transmitters may be added to this network at a later date to fill in the remaining gaps in the frequency and time sequence.

VHF and UHF wave propagation

For frequencies above about 50 MHz the sky wave is not normally reflected by the ionosphere, even during a sunspot maximum period. Here the normal mode of propagation follows a simple line-of-sight path which is limited by the curvature of the earth's surface to the effective horizon distance. In practice the signal does extend some distance beyond the physical horizon but beyond this point the normal signal drops off very rapidly.

Radio horizon distance

For line-of-sight propagation the distance is limited by the curvature of the earth. The effective horizon distance for a VHF radio wave is slightly greater than the visual horizon distance because the wave path tends to be bent slightly by the atmosphere. By assuming an effective earth radius which is 1.33 times its actual value, the radio horizon distance is given by:

$D = 1.42H$

where D = horizon distance in miles
        H = antenna height above sea level in feet

Tropospheric propagation

Long-distance communication on VHF and UHF is possible due to propagation effects within the dense lower region of the atmosphere, called the troposphere. An effect known as tropospheric 'ducting' can occur, when a temperature inversion is present along the path between the two stations. Normally the temperature of the atmosphere falls with increasing height. In temperature inversion an upper layer of the atmosphere is at a higher temperature than the layer at ground level, and this causes refraction of the VHF wave (Figure 11.4).

Tropospheric ducting can give propagation over hundreds of miles on 144 and 432 MHz. The effect usually occurs under high barometric pressure conditions in the early evening after a warm dry day, particularly when there is fog or moist conditions at ground level. This mode tends to occur in the summer and autumn.

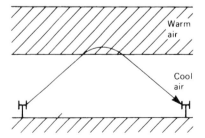

Figure 11.4 Tropospheric ducting of VHF signals produced by a temperature inversion in the lower part of the atmosphere

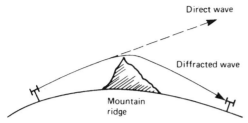

Figure 11.5 Propagation of VHF signals over a mountain ridge by diffraction of the wavefront

Wave diffraction effects

A ridge of hills or mountains can produce noticeable diffraction effects upon the VHF wave so that it is effectively bent to give reception beyond the ridge. The effect depends upon the angle at which the wave travels to the ridge and the sharpness of the edge of the ridge (Figure 11.5).

Sporadic E propagation

From time to time the E layer can reflect signals in the 25–80 MHz range over long distances. It is thought that this sporadic-E-type propagation involves the ducting of the signal through the E layer before it returns to earth. This type of propagation is of interest to amateurs operating in the 50- and 70-MHz bands, when contacts over thousands of miles may be possible.

Auroral propagation

An aurora is an electromagnetic storm occurring in the polar regions and accompanied by luminous displays in the sky. On the HF bands an aurora will usually cause disruption of communications due to severe static and fading problems. The auroral display causes heavy ionization of the atmosphere which can provide a reflective area for VHF and UHF signals, allowing communications over long paths (Figure 11.6). The basic technique is that stations aim their beam antennas at the aurora, from which the signal is reflected to the distant station. Since the ionization within the aurora changes rapidly, auroral contacts are accompanied by fading and multipath distortion, but can provide useful communications on 50 and 144 MHz using CW or SSB.

Amateur VHF and UHF beacons

A series of beacon transmitters has been set up on the 50-MHz band to provide an indication of the presence of sporadic E or normal F layer

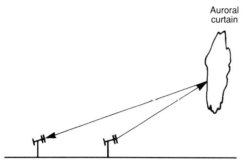

Figure 11.6 Reflection of VHF signals from the ionized curtain of an auroral storm

propagation on this band. If a beacon can be heard, this indicates that a propagation path exists between the receiving station and the region where the beacon is located.

As on 28 MHz, there are a large number of beacons on 50 MHz and their status changes from time to time. A selection of the 50-MHz beacons is given here as a guide.

On the higher frequency VHF and UHF bands, beacons have also been set up to allow amateurs to detect any lift conditions that might occur which would permit Dx contacts on these bands.

Six-metre band (50 MHz)

Call	Frequency (MHz)	Location
GB3BUX	50.000	Buxton, England
H44HIR	50.005	Solomon Islands
FY7THF	50.039	French Guiana
SZ2DH	50.015	Athens, Greece
GB3SIX	50.020	North Wales
6Y5RC	50.025	Jamaica
ZB2VHF	50.035	Gibraltar
OX3VHF	50.045	Greenland
GB3NHQ	50.050	London, England
ZS6LN	50.055	Transvaal, RSA
GB3RMK	50.060	Inverness, Scotland
PY2AA	50.062	Sao Paulo, Brazil
W3VD	50.062	Laurel, Md. USA
GB3LER	50.064	Lerwick, Scotland
EA3VHF	50.070	Barcelona, Spain
VS6HK	50.075	Hong Kong
TI2NA	50.080	Costa Rica
9H1SIX	50.085	Valletta, Malta
VE1SIX	50.088	New Brunswick, Canada
VK8VF	52.200	Darwin, Australia

Four-metre band (70 MHz)

Call	Frequency (MHz)	Location
GB3BUX	70.000	Buxton, Derbyshire
GB3CTC	70.030	St Austell, Cornwall
GB3REB	70.040	Chatham, Kent
GB3ANG	70.060	Dundee, Scotland
5B4CY	70.110	Cyprus
ZB2VHF	70.120	Gibraltar
EI4RF	70.130	Slane, Eire

Two-metre band (144 MHz)

Call	Frequency (MHz)	Location
DL0UB	144.850	Trebbin, Germany
HB9H	144.850	Locarno, Switzerland
LA5VHF	144.855	Bodo, Norway
LA1VHF	144.860	Rjukan, Norway
HB9HB	144.865	Biel, Switzerland
LA2VHF	144.870	Storen, Norway
HB9W	144.875	Zurich, Switzerland
LA3VHF	144.880	Mandal, Norway
GB3MCB	144.915	St Austell, Cornwall
E12WRB	144.920	Portlaw, Eire
GB3VHF	144.925	Wrotham, England
GB3LER	144.965	Lerwick, Scotland
GB3ANG	144.975	Dundee, Scotland
ON4VHF	144.985	Belgium

70-centimetre band (432 MHz)

Call	Frequency (MHz)	Location
GB3WHA	432.810	Kent
EI2WRA	432.870	Portlaw, Eire
GB3SUT	432.890	Sutton Coldfield, England
GB3MLY	432.910	Elmley Moor, England
GB3LER	432.965	Lerwick, Scotland
GB3MCB	432.970	St Austell, Cornwall
GB3ANG	432.980	Dundee, Scotland

23-centimetre band (1290 MHz)

Call	Frequency (MHZ)	Location
GB3NWK	1296.810	Orpington, Kent
GB3MHL	1296.830	Martlesham
GB3FRS	1296.850	Farnborough, Hampshire
GB3MCB	1296.860	St Austell, Cornwall
GB3DUN	1296.890	Dunstable, Bedfordshire
GB3IOW	1296.900	Isle of Wight
GB3CLE	1296.910	Clee Hill, Salop
GB3MLE	1296.930	Emley Moor, England
GB3ESB	1296.970	Hastings, Sussex
GB3EDN	1296.990	Edinburgh, Scotland

13-centimetre band (2300 MHz)

Call	Frequency (MHz)	Location
GB3MHS	2320.830	Martlesham
GB3NWK	2320.850	Orpington, Kent
GB3BSY	2320.880	Barnsley
GB3ANT	2320.890	Norwich, Norfolk
GB3WWH	2320.910	Westbury, Wiltshire
GB3LES	2320.955	Leicester

Propagation bulletins

Broadcast bulletins of current and forecast propagation conditions are transmitted as follows:

WWV, Fort Collins, Colorado, USA
Frequencies: 2500, 5000, 10 000, 20 000 and 25 000 MHz

Bulletin transmitted at 15 min past each hour. Data given are the solar flux, A index, Boulder K index, solar activity and geomagnetic field conditions for the past 24 hours and predicted values for the next 24 hours. Announcement is in voice using AM.

Propagation conditions are also included in the ARRL News, broadcast daily by W1AW, and in the RSGB News, broadcast on Sundays by GB2RS.

Solar flux

This is a measure of the radiation from the sun at 2800 MHz. A reading of 60 represents a 'quiet' state. With readings over about 80, the higher HF bands (21 and 28 MHz) will open up for Dx. Readings above 200 give 24-hour operation on 21 and 28 MHz and useful openings on 50 MHz.

A and K propagation indices

This is a measure of the earth's geomagnetic field. The A index is for a period of 24 hours and has a scale from 0 to 400. The K index is similar but uses a different scale and is updated more frequently. Low values of A or K index give best HF propagation. Higher values may indicate auroral conditions which may be useful for VHF operations.

Solar activity

This is an indication of rapid changes in solar flux and solar storms. It is indicated as low, medium, high or very high. High activity indicates propagation disturbances, particularly on the HF bands.

Geomagnetic field conditions

These are usually quoted as quiet ($K < 1$), unsettled ($K = 1$–3) or active $K > 3$). Best conditions are when the state is quiet, since unsettled or active conditions indicate magnetic storms and probable HF blackout.

Meteor scatter propagation

One interesting mode of propagation that has been used by amateur stations is reflection from the ionized trail produced by a shower of meteors. During the year there are a number of showers of meteors which can provide meteor scatter contacts. Prominent showers are the Perseids in August and the Geminids in December. This mode of propagation may be used on the 50- and 144-MHz bands but can provide only brief contacts, since the meteor trail will exist only for a short period of time.

Meteor scatter propagation is a rather specialized mode and contacts are usually made using a timing sequence for transmitting and receiving. For CW, each period is 5 min long, and stations transmit for one 5-min period and then listen during the next 5-min period. Stations beaming north or west transmit during the first 5-min period after each hour, whilst those beaming south or east transmit during the second 5 min. When using SSB the same procedure is used but the timing periods are reduced to 1 min in duration. Morse transmissions may be made at speeds of 80 to 100 words per minute by using automatic keying systems or computer control.

Computer programs can be used to predict the time and direction of the meteor showers so that the antenna can be aimed in the correct direction. Some of the major meteor showers are:

Quadrantids	1–5 January
Arietids	1–16 June
Perseids	20 July to 18 August
Geminids	7–15 December

12

Antennas

An important part of any radio station is the antenna, which when driven by RF power from a transmitter will generate the electromagnetic fields which will propagate the signals to a remote receiver. At the receiving end the electromagnetic waves interact with the antenna to produce RF voltage at the antenna terminals which is then fed to the input of the receiver.

Antennas can have a variety of forms, ranging from simple single-element types to complex arrays of elements. All antennas have a number of common parameters which will determine the performance of the antenna and how it matches to the output of the transmitter or the input of the receiver.

Antenna impedance

All antennas exhibit an impedance at their feed point. This may consist of a resistance and a reactive component which looks like either inductance or capacitance. Ideally the antenna impedance should be purely resistive.

In a resonant antenna, such as a dipole, the impedance will be a pure resistance at the resonant frequency of the antenna. For non-resonant antennas the impedance contains a capacitive or inductive component as well as resistance.

For optimum transfer of power from the transmitter to the antenna, the output impedance of the transmitter should be matched to the impedance of the antenna that it is driving. Transmitter output impedance for a modern transceiver is generally arranged to be 50 Ω and is purely resistive. If the antenna also shows a resistance of 50 Ω, then the two may be connected together and maximum power will be transferred between transmitter and antenna.

In many cases the antenna may not have a 50-Ω impedance, and a matching circuit, known generally as the antenna tuning unit or ATU, is inserted between the transmitter and the antenna. The ATU is used to cancel out any reactive components of the antenna impedance and to match the resistive component to the 50-Ω transmitter output.

Antenna gain

The effective gain of the antenna is the ratio between the radiated power in the direction of maximum radiation of the antenna and the radiated power from an isotropic radiator fed with the same input power. An isotropic radiator is a theoretical antenna which radiates equally in all directions. The gain of an antenna is normally given in decibels. As an example, the gain of a half-wave dipole is 2.14 dB.

In some cases the antenna gain may be referred to that of a half-wave dipole, since it is much easier to make the measurements if a dipole is used as the reference. This gives an antenna gain figures which is about 2 dB lower than the calculated gain figure based upon an isotropic radiator.

Polar diagrams

Virtually all types of practical antenna do not radiate equally in all directions. The polar diagram for an antenna shows the radiation pattern of the antenna in one plane. The polar diagram which is usually of interest is the plan view, which shows the pattern of radiation around the antenna in the horizontal plane (Figure 12.1).

For a directional beam antenna the radiation is primarily in one direction and there is virtually no radiation from the sides and back of the antenna. Any small areas of signal at the sides and back of the radiation pattern are referred to as side lobes. In the design of a directional antenna, the object is to reduce these to the smallest possible size.

The beam-width is the angular width of the main lobe at the point where the radiation power is reduced by 3 dB relative to the maximum power point at the centre of the main lobe.

The ratio between the power in the forward direction of the antenna and that from the rear is referred to as the 'front-to-back ratio' of the antenna. Antenna design is usually a compromise between achieving high forward gain with a narrow beamwidth and retaining a high front-to-back ratio.

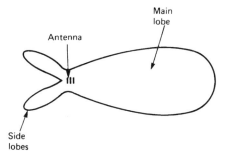

Figure 12.1 Polar diagram showing the directional properties of an antenna

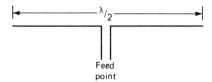

Figure 12.2 A half-wave dipole or doublet antenna

A second polar diagram may be produced which gives the radiation pattern in the vertical plane.

The dipole antenna

Perhaps the simplest basic antenna is the half-wave dipole, which is sometimes called a doublet antenna. The dipole consists of a wire which is approximately half a wavelength long at the frequency of operation. The wire is broken at the centre and signals are fed to or from the antenna via a two-wire feeder cable as shown in Figure 12.2.

When the dipole is operated at its design frequency, the impedance at the centre feed point is approximately 70 Ω and because the antenna is resonant at that frequency, there is no reactance component. This assumes that the dipole antenna is mounted at least one wavelength away from ground or other solid objects, such as buildings. On VHF and UHF bands this is not difficult to achieve, since the antenna is small and wavelengths are short. On the HF bands the antenna may well be mounted less than a wavelength above ground, and the effect is to reduce the impedance of the antenna and to alter its radiation pattern, since the ground tends to act as a reflector.

A practical dipole antenna has a length which is approximately 5% less than the calculated value for half a wavelength at the frequency where it is to be used. A useful formula for calculating the antenna length is:

$$\frac{143}{f(\text{MHz})} \quad \text{metres}$$

When operating as a transmitting antenna the dipole will have a current distribution which gives maximum current at the centre feed point and minimum current at the ends of the antenna. The voltage will be low at the feed point and high at the ends. For a transmitter running high power this can mean that there are very high voltage levels at the ends of the wire.

The radiation pattern for a horizontally mounted dipole is a figure-of-eight pattern giving maximum signal at right angles to the dipole and minimum signal from the ends of the dipole (Figure 12.3). If the antenna is mounted with the wire running vertically, then the radiation pattern is equal in all directions in the horizontal plane (Figure 12.4).

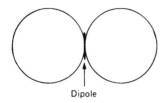

Figure 12.3 Polar diagram for a horizontal dipole

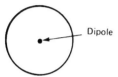

Figure 12.4 Polar diagram for a vertical dipole

Multi-band dipoles

It is possible to produce a dipole-type antenna which will operate on a number of different bands by joining two or more dipole antennas in parallel. This relies upon the fact that at resonance a dipole has a low resistance whilst on frequencies well away from resonance the impedance is high. Thus three dipoles cut for 3.5, 7, 14 and 28 MHz could be connected in parallel at the feed point. To reduce inductive coupling between the antennas, the ends of the three dipoles may be spaced apart as shown in Figure 12.5. When driven by a 7-MHz signal, only the 7-MHz dipole will be active, since the other two show a high impedance. The dipoles will also show a low impedance at their third harmonic frequency, so that the 3.5-MHz dipole will work quite well when operated at 10 MHz, and the 7-MHz dipole will radiate quite effectively on 21 MHz. Connecting a dipole for 10 MHz in parallel with a 3.5-MHz dipole will not work very well, because both antennas have a low impedance at 10 MHz.

With a parallel dipole-type antenna the other dipoles will have some effect upon the tuning of the higher frequency dipoles. Cut the higher

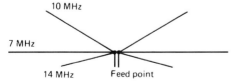

Figure 12.5 A multi-band antenna using three dipoles which are connected in parallel

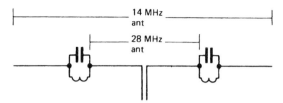

Figure 12.6 A multi-band dipole antenna using tuned traps to isolate sections of the antenna according to the band in use

frequency dipoles so that they are slightly too long. Adjust the length of the lowest frequency dipole for resonance and a proper match to the transmitter. Then adjust the lengths of the higher frequency antennas in turn until a good match is achieved on each band. A reasonable match should also be present on the third harmonic bands of the lower frequency dipoles.

Trap dipole

An alternative technique for making a dipole antenna operate on several bands is to insert a pair of tuned traps into a low-frequency dipole as shown in Figure 12.6. The traps are tuned circuits which act as rejectors for the higher frequency band and effectively isolate the outer sections of the dipole when the higher band is being used.

Suppose the higher frequency band is to be 28 MHz and the lower band 7 MHz. The centre section of the antenna, between the traps, is cut for operation on 28 MHz and the overall length is cut for 7 MHz. The traps are tuned to resonate at 28 MHz so that when the antenna is used on 28 MHz only the centre portion is active. When the antenna is driven at 7 MHz the traps are not resonant and will appear as series inductors, so the whole antenna works as a 7-MHz dipole. Because of the traps, the centre portion will need to be shorter than a normal 28-MHz dipole and the overall length will be shorter than for a 7-MHz dipole. These lengths will need to be adjusted by trial and error for best results on the two bands. By adding more sets of traps the antenna could be made to operate on five or six different bands, which is useful for those amateurs who do not have space to erect large arrays of antennas.

Marconi antenna

An alternative to the dipole antenna is the Marconi type, in which a quarter-wave long vertical wire is fed at the bottom with the earth as its second connection as shown in Figure 12.7. Often the antenna is simply a vertical tower which is mounted on an insulated base. The Marconi produces vertically polarized signals and has a typical impedance of about 30–35 Ω. The

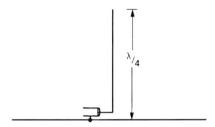

Figure 12.7 Marconi quarter-wave vertical antenna

Marconi gives equal all-round radiation in the horizontal plane. In the vertical plane it produces a lobe at an angle of about 30°.

Ground plane antenna

The ground plane antenna is basically a quarter-wave vertical Marconi-type antenna with its own ground plane screen. In practice the ground plane often consists of just four radial quarter-wave elements, as shown in Figure 12.8. Thus although the antenna may be mounted high above actual ground, it operates against its own ground plane. If the antenna is mounted at ground level, the radials may be buried a few inches underground and there will usually be more of them.

This antenna has the advantage that it produces a low angle of radiation in the vertical plane when compared with other types of antenna. Typically the ground plane has a main lobe at around 30° above horizontal, which makes it an ideal antenna for Dx operation on the HF bands (Figure 12.9). In the horizontal plane the ground plane antenna is omnidirectional. Feed impedance is nominally about 35 Ω and the feed is unbalanced, so the ground plane antenna is normally fed by a coaxial feeder cable.

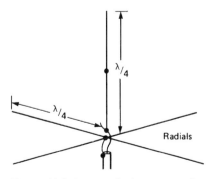

Figure 12.8 A ground plane type of antenna using four quarter-wave radials as a ground plane

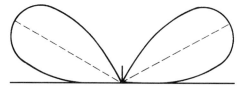

Figure 12.9 Radiation pattern in the vertical plane for a Marconi or ground plane type of antenna

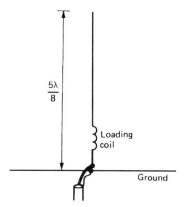

Figure 12.10 A 5/8 wavelength vertical antenna with loading coil at the bottom for impedance matching and resonating the antenna

Loaded vertical antennas

Often it is not practical to use a quarter-wave vertical antenna, so a shorter antenna is used. In order to resonate the antenna and obtain a reasonable impedance for matching, a series inductor may be added. In most cases this loading coil is mounted at the base of the antenna but a better position is either at the middle or top of the antenna (Figure 12.10).

Loaded vertical antennas are frequently used as mobile antennas and the metal body of the vehicle then acts as a ground plane for the antenna. In some cases longer antennas of 5/8 wavelength are used. Here the upper part of the antenna acts as an end-fed half-wave dipole and the lower part as a loaded quarter-wave vertical.

For handheld units a helical antenna is used. This is an extreme example of loading in which the loading coil becomes the antenna.

Antenna polarization

If a wire antenna such as a dipole is mounted horizontally, the electric field of the radiated wave is in the horizontal plane and the magnetic field is vertical. This produces a horizontally polarized wave.

A vertically polarized wave is one in which the electric component of the wave field is in the vertical plane and the magnetic component is the horizontal plane. This type of polarization is produced by a vertical wire antenna such as a Marconi or ground plane type.

An antenna which is sloping produces a mixture of both horizontal and vertically polarized waves.

It is possible to produce circularly polarized waves by feeding crossed antennas (one horizontal and one vertical) with signals that are 90° out of phase. This type of polarization is often used at VHF and UHF and for satellite antennas.

Beam antennas

By combining together a number of simple antennas, such as dipoles, it is possible to produce a very directional antenna which also exhibits considerable gain in signal over a single dipole in the direction of maximum radiation or pickup. These beam antennas fall into two basic groups. In the first group the antennas are all driven from the transmitter and the signals to each antenna in the array are carefully phased relative to those in the other antennas so that the signal is reinforced in the main beam direction and cancelled in the other directions. The second type of beam antenna uses parasitic elements which are not fed directly by the transmitter but which re-radiate the signal that they themselves pick up from the driven antenna element.

Yagi-type beam antennas

In the Yagi-type antenna a reflector element is mounted behind the dipole and one or more director elements may be mounted in front of the dipole (Figure 12.11). The reflector is a short-circuited dipole which is made slightly longer

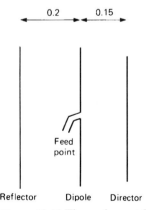

Figure 12.11 Three-element Yagi-type beam antenna

than half a wavelength. The effect of detuning the reflector is to alter the phase
of the current induced in it by the signal radiated from the dipole. The current
flowing in the reflector causes it to radiate, and if the phasing is correct this
radiated signal will reinforce that of the dipole in one direction and cancel it
in the other. The result is that radiation behind the reflector is reduced to give
a directional beam in the direction of the dipole. The reflector is called a
parasitic element since it is not driven directly from the transmitter.

If the parasitic element is shorter than the driven element it becomes a
director and the radiation is increased in the direction of the director. A
typical Yagi beam antenna will have a reflector and one or more directors.
It was found that adding extra reflector elements had little effect.

The optimum spacing of the reflector and directors is normally less than
a quarter wave, and typically the reflector is spaced at about 0.2 wavelength
behind the dipole and the directors at 0.15 wavelength intervals in front of
the dipole. As the number of elements in the antenna rises, the forward gain
increases and the beamwidth becomes narrower. Antenna with up to 18
elements are not unusual, but beyond this point adding extra directors has
less effect.

For higher antenna gain, Yagi arrays can be stacked vertically or horizontally or both ways. The feed cables between the individual Yagi arrays must
be arranged so that the arrays are fed in the correct phase relative to one
another so that the overall radiation pattern is reinforced.

Transmission lines

In most transmitting stations the antenna is at some distance from the transmitter itself and some means of conveying the signal efficiently between
transmitter and antenna is required. This link is provided by an antenna
feeder which may be a cable or an open wire arrangement.

The transmission line has a characteristic impedance Z_o. If the line is terminated in a resistive load equal to Z_o, it is correctly matched and the maximum
amount of power is transferred along the line to the load. Under these conditions the load presented to the transmitter by the matched line is a pure resistance equal to Z_o

Coaxial cable

For most applications the transmission line is a coaxial cable. This has a
centre conductor and a concentric outer conductor (screen) which are
separated by insulation (Figure 12.12). The characteristic impedance of the
cable is determined by the geometry of the centre conductor and the outer
screen and is given by:

$$Z_o = 138 \times \log \frac{D_o}{D_i} \ \Omega$$

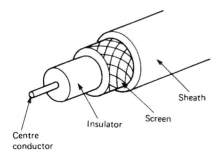

Figure 12.12 Construction of a coaxial feeder cable

where D_i is the diameter of the inner conductor, and D_o is the diameter of the outer conductor. Typical coaxial cables have a characteristic impedance of either 50 Ω or 75 Ω.

Losses in a coaxial cable depend upon the conductivity of the centre conductor and screen and the characteristics of the insulation between them. In low-loss cables the insulator is made in the form of a helix so that most of the insulation is provided by air rather than solid plastic. In some cases the centre conductor may be silver plated or even solid silver to reduce losses. Attenuation is quoted in decibels per 10-m length of cable. Total power loss over a length of cable is given by:

$$\text{Total loss (dB)} = \frac{\text{Length (m)} \times \text{Attenuation (dB/10 m)}}{10}$$

The characteristics of some typical coaxial cables are as follows:

Type	Diameter (mm)	Z_o (Ω)	pF/m	Attenuation (dB/10 m)	
				10 MHz	100 MHz
RG8	10.2	50	96	0.3	0.9
RG11	10.2	75	66	0.3	0.9
RG58C	5.0	50	100	–	2.0
RG59B	6.15	75	68	–	1.3
RG174A	2.8	50	100	1.1	2.8
RG178B	1.8	50	96	1.8	4.4
RG179B	2.5	75	64	1.9	3.2
UR43	5.0	50	100	0.7	1.3
UR67	10.3	50	100	–	0.7
UR70	5.8	75	67	–	1.5
UR76	5.0	50	100	–	1.6
UR95	2.3	50	100	0.9	2.7

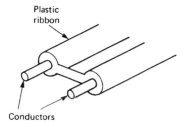

Plastic
ribbon

Conductors

Figure 12.13 Construction of a flat twin cable

Twin-wire feeders

An alternative to coaxial cable is the twin-wire feeder, where two parallel wires are embedded in a flat ribbon cable (Figure 12.13). Standard cables are made with either 75-Ω or 300-Ω impedance and their characteristics are:

Z_o (Ω)	Width (mm)	pF/m	Attenuation (dB/10 m)	
			10 MHz	1000 MHz
75	4.0	60	0.12	2.0
300	9.7	13	0.12	1.7

An air-spaced twin feeder can be made by having two parallel wires held apart by insulating spacers at intervals along the line. The characteristic impedance of an air-spaced parallel twin line is given by:

$$Z_o = 276 \times \log \frac{2S}{d}$$

where S is the spacing between the wires and d is the conductor diameter.

Tuned wire feeders

An alternative to the conventional twin feeder is the use of a widely spaced pair of feeder wires which are tuned as part of the antenna system itself. In this case there is normally a standing wave on the feeders themselves but they do not radiate because the currents in the two wires are in opposite directions and the fields tend to cancel.

The feeders and the antenna system are tuned to resonance so that the impedance seen at the transmitter end is non-reactive and the antenna matching and tuning unit is then used to transform this impedance to match the output impedance of the power output stage of the transmitter.

Wave velocity factor

Radio waves travelling along a feeder cable have a velocity which is less than that of a wave in free space. The ratio between the velocity in a cable and that in free space is called the velocity factor and needs to be taken into account when cables are used to link several antennas operating as an array, since the length of the link cables will determine the phasing of the feeds to the individual radiator elements.

Voltage standing wave ratio (VSWR)

When a transmission line is mismatched at one end, some of the transmitted signal is reflected back along the line. This reflected signal varies in phase with the forward signal as it travels back along the line, and the effect is that it will either add to or subtract from the forward signal, thus producing a variation in the effective signal according to the position along the line. This produces what is known as a standing wave along the transmission line.

A convenient method of measuring the degree of mismatch between the line impedance and the load at the end of the line is that of measuring the relation between the forward signal and the reflected signal on the line. This measurement is known as the voltage standing wave ratio (VSWR), and is often referred to simply as the SWR.

In an ideally matched line the VSWR is $1 : 1$. On the HF bands an SWR of less than $1.5 : 1$ is acceptable, since the cable losses on HF are not high and even a solid-state output amplifier can usually handle this level of SWR. On VHF and UHF, where cable losses can be high, a low SWR can be important and an SWR of better than $1.2 : 1$ is desirable.

Balun transformers

A symmetrical antenna such as a dipole will produce signals at its two terminals which are balanced relative to ground potential, whereas a vertical antenna such as a ground plane provides a signal where one side of the signal source is at ground potential. Most transmitters and receivers have an unbalanced output or input in which one side of the signal is tied to ground potential. For optimum performance the dipole should be fed from a balanced signal source and this can be achieved by including a balun (balanced to

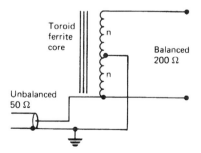

Figure 12.14 Balun transformer which gives a $4 : 1$ impedance ratio

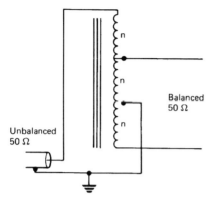

Figure 12.15 Balun transformer which gives a 1 : 1 impedance ratio

unbalanced) transformer in the transmission line system. On the balanced side the centre tap point is grounded and the two feeds to the antenna operate in antiphase to give a balanced feed to the dipole or other balanced input antenna. The transmitter side of balun is grounded on one side to match the unbalanced output of the transmitter.

If the windings are arranged to have equal turns, the transformer provides an impedance transformation and the impedance at the balanced side will be four times that at the unbalanced side (Figure 12.14). If a 1 : 1 ratio is required, then an extra winding is added as shown in Figure 12.15.

13

Transmitters

The basic block diagram for a radio transmitter is shown in Figure 13.1. The first requirement is to generate a stable and accurate carrier frequency and this is the function of the oscillator section. This block may consist of a simple oscillator or may involve more complex signal-generating systems such as a digital frequency synthesizer.

Once a carrier signal has been produced, it is passed through a power amplifier which raises the power level to feed the transmitting antenna. The actual output power required may vary from one or two watts for a handheld VHF unit through 100 W for an amateur HF transmitter to hundreds of kilowatts for a broadcast transmitter.

In order to convey information, the carrier signal from the transmitter needs to be modulated by some form of message signal. Early transmissions simply switch the carrier on and off to provide a sequence of dots and dashes comprising the Morse code. Most transmissions use an audio signal, such as speech, to modulate the amplitude of the carrier signal, and here the modulator section usually varies the supply voltage or current to the final power amplifier stage. In an SSB transmitter the modulation is generated in the low-power driver stages of the transmitter. For other modulation techniques, such as FM, the modulator controls the oscillator section of the transmitter.

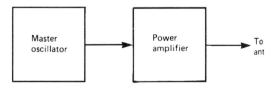

Figure 13.1 Block diagram of basic transmitter

Crystal oscillators

The simplest form of frequency generator for a transmitter is a crystal-controlled oscillator which uses a quartz piezoelectric crystal to control the frequency of oscillation. The quartz crystal consists of a thin slab of quartz mounted between two metal electrodes. When excited by an electrical signal the quartz plate vibrates mechanically and the frequency of oscillation is governed by the physical dimensions of the quartz plate. When the crystal vibrates it also generates a small electrical signal at the vibration frequency. If this signal is amplified and then fed back so that it reinforces the mechanical oscillation of the crystal, it is possible to produce an oscillator with a very stable output frequency.

There are a number of different ways in which the slab of quartz can be cut from the original quartz crystal and these lead to different characteristics in terms of frequency stability and ease of obtaining oscillations. A quartz crystal can operate in either the series resonant or parallel resonant modes. In the series mode the impedance is relatively low at perhaps 10 000 Ω, whilst in the parallel mode the crystal shows an impedance of several megaohms. The frequency of oscillation is slightly higher when the crystal is operating in the parallel mode. The capacitance of the holder may also affect the frequency slightly.

The advantage of using a quartz crystal instead of a conventional LC tuned circuit is that the crystal has an effective Q of several thousand, compared with perhaps 100 for an LC circuit, and the frequency produced is very stable.

A typical crystal oscillator circuit is shown in Figure 13.2. This is a Colpitts-type circuit in which the crystal acts as a tuned filter in the feedback path so

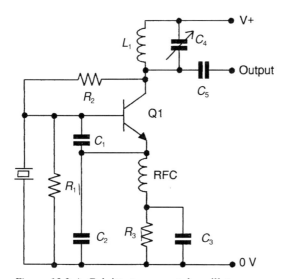

Figure 13.2 A Colpitts-type crystal oscillator

that oscillation occurs at the resonant frequency of the crystal. In fact the oscillator will also produce harmonics of the crystal frequency, so a tuned filter after the oscillator is desired to remove the harmonic components from the signal.

Quartz crystals for operation in their fundamental mode are available for frequencies from a few hundred kilohertz to perhaps 12 MHz. For higher frequencies the crystal plate becomes very thin and more difficult to manufacture. For signal generation at these higher frequencies the crystals are designed to operate in the harmonic or overtone mode where the actual frequency of oscillation is at a harmonic of the fundamental oscillation frequency of the quartz slab. In oscillators operating in this mode the output tuned circuit of the oscillator is resonated at the desired harmonic frequency. This technique can be used to provide crystal-controlled oscillators which will generate stable frequencies into the lower VHF range.

Variable-frequency oscillators

For applications such as amateur radio, it is desirable to have a transmitter which can readily be tuned to a number of different frequencies within a band. This can be achieved, to a limited extent, by using a crystal oscillator and switching in different crystals to give a selection of different frequencies. This arrangement was often used on early VHF transmitters. The ideal arrangement is to have an oscillator that can be tuned across a frequency band to any desired frequency.

The simplest variable-frequency oscillator (VFO) uses an LC tuned circuit to determine the frequency, and a variable capacitor is used to select a particular frequency within the operating band. The major problem with this type of oscillator is that of keeping the frequency stable and determining the exact frequency of operation.

For amateur radio use, the VFO is usually designed to operate at a relatively low frequency such as 1.8 or 3.5 MHz. Higher frequencies are then obtained by using frequency multiplier circuits, as shown in Figure 13.3. The frequency multiplier stages are simply amplifiers which are operated so that they generate strong harmonics of the input frequency, and a tuned circuit at the output selects either the second or third harmonic to provide frequency doubling or frequency trebling.

Figure 13.3 Using a frequency doubler with a VFO to obtain higher output frequencies

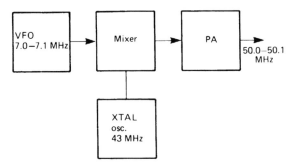

Figure 13.4 System which mixes a VFO signal with that from a crystal oscillator to achieve stable high-frequency signals

An alternative technique is to mix the VFO signal with a signal from a crystal oscillator. This produces sum and difference frequencies at the output of the mixer stage. A tuned circuit is then used to select the sum frequency at the output of the mixer circuit. This arrangement is shown in Figure 13.4.

Today the VFO has tended to fall into disfavour, since the development of digital frequency synthesizers has made the generation of a wide range of accurate and stable frequencies much easier.

Digital synthesizer

The basic arrangement of a simple frequency synthesizer is shown in Figure 13.5. Here a voltage controlled oscillator (VCO) is used as the VFO. In this type of oscillator a control voltage applied to the oscillator governs the frequency of operation. This may be achieved by using a variable-capacitance diode as the tuning capacitance in the oscillator tuned circuit which determines oscillator frequency. When the DC voltage applied to the diode is varied, its capacitance changes and alters the frequency of the oscillator. The output of the oscillator is fed to a frequency divider and the divided-down frequency is then compared in a phase director with the frequency of a reference oscillator. The output of the phase detector is then fed through a low-pass filter to produce the control voltage for the VCO. The frequency divider is programmable so that the division ratio can be set to any value n. When the circuit is operating properly, the VCO will be operating at n times the frequency of the reference oscillator. If the reference oscillator were set at, say, 10 KHz, then the VCO frequency could be set at 10-kHz intervals through a band of frequencies. Normally the VCO will only operate correctly over a limited range of frequencies, but if the divider is set to give ratios from 1400 to 1435 the VCO would cover the 14-MHz band in 10-kHz steps.

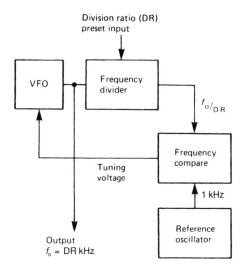

Figure 13.5 Block diagram of a digital frequency synthesizer

Actual synthesizer circuits use special integrated circuits to perform most of the synthesizer functions. A typical synthesizer system used in an amateur transceiver will allow the frequency to be adjusted in steps of about 100 Hz to give very precise control of the output frequency.

Classes of operation

The power amplifier stages used to increase the power level to the desired output power may have different classes of operation according to the way in which the amplifer stage is biased. The class of operation is defined by the angle of the signal cycle over which the amplifier devices (valves or transistors) are conducting.

A class A amplifier is one where current flows throughout the entire signal cycle, giving a conduction angle of 360°. This type of amplifier is linear in operation so that the output waveform is the same as the input and minimum distortion of the signal occurs. This is the type of operation used for most audio amplifier stages. A typical class A power amplifier will have a maximum efficiency of 50%, and typical amplifiers usually have efficiencies of around 40–45%. Thus the output power is about half the input power, and the output transistors or valves have to dissipate the remaining half of the input power as heat.

Class B operation results when the amplifier devices conduct for one half-cycle (180°). If the amplifier is a single valve or transistor it will produce pulses of current at its output. By placing a tuned circuit at the output the signal can be restored to a sinewave form, since the pulses of current from the amplifier will excite the tuned circuit into oscillation so that it generates the second half-cycle of the output signal.

In a class B amplifier the efficiency can be up to 63%, so that for the same output power the transistor or valves have to dissipate less power as heat. When used as an RF amplifier driving a tuned output circuit this type of amplifier is linear in operation, so that a modulated signal applied at the input will be amplified with minimal distortion.

For audio amplifiers operated in class B, two transistors or valves are used in a push–pull arrangement where one transistor, or valve, conducts on one half-cycle and the other on the second half-cycle. Thus power is delivered to the output load on both half-cycles and the circuit produces linear amplification of the input signal. This arrangement is widely used for audio power output stages and can also be used in RF power amplifiers for transmitters.

If the amplifier devices conduct for less than a half-cycle, the amplifier becomes a class C type. Typically, a class C amplifier has a conduction angle between 120° and 160°. The amplifier stage produces a short pulse of current during each cycle of the input signal which excites the output tuned circuit into oscillation to produce a sinewave output signal. Because the amplifier stage conducts for only part of the signal cycle, the efficiency of a class C amplifier can be very high, with typical amplifiers giving 60–80% efficiency. Class C amplifiers are non-linear in operation and cannot be used to amplify AM or SSB signals. This type of amplifier can be used in CW or FM transmitters where the modulation is carried out in an early stage of the transmitter. For an AM transmitter, class C stages can be used, provided that the amplitude modulation is applied to the final amplifier stage.

Amplitude modulation (A3E)

In amplitude modulation the RF signal amplitude is varied in sympathy with a modulating waveform such as a speech signal. The resultant signal has a spectrum which consists of a constant-amplitude carrier component plus two sideband signals, one on each side of the carrier. The sidebands correspond to the sum and difference components of the carrier and the modulating frequency. Thus a 1-kHz tone modulating a 1-MHz carrier produces sideband frequencies of 999 kHz and 1001 kHz.

Signal bandwidth for A3E is twice the highest modulating frequency. For typical voice signals of 300 Hz to 3 kHz the bandwidth required is 6 kHz. For a broadcast transmitter a bandwidth of around 20 kHz would be typical in order to provide reasonable audio quality.

AM is used by all medium-wave broadcast stations and almost all short-wave broadcast stations. Although this mode was widely used on the amateur bands until the 1960s, it has fallen into disuse and very few amateurs now use AM for speech contacts. AM is widely used on the VHF aircraft bands and for some mobile radio stations. This type of modulation is also used for the video signal on television transmissions.

A simple form of AM is the keyed carrier technique used to send Morse code. This has the emission designation A1A for Morse or A1B for an RTTY-type signal. Another method is to key an audio tone which is then used to modulate the transmitter. This has the emission designation A2A for Morse or A2B for RTTY.

In an AM transmitter the modulation is usually applied to the final RF amplifier stage by modulating its power supply feed. If the amplitude modulation is carried out in one of the low power stages of the transmitter then all of the following power amplifiers must operate in a linear mode using either class A or class B amplifier stages.

Frequency modulation (F3E)

In frequency modulation, as its name implies, the frequency of the carrier wave is altered in sympathy with the amplitude variations of the modulating signal. The amount by which the carrier frequency is changed by the modulation is called the 'frequency deviation' and is proportional to the amplitude of the modulating signal. The RF output power remains constant during the modulation cycle.

The bandwidth of the FM signal is given approximately by:

$$B = 2(M + D)$$

where

B = bandwidth (Hz)
M = highest modulation frequency (Hz)
D = peak deviation (Hz)

Amateur stations use narrow-band FM on the VHF bands with a peak deviation of 5 kHz. Assuming that the audio bandwidth for speech is limited to a highest frequency of 3 kHz, the bandwidth of these signals is approximately 16 kHz. On the HF bands amateur FM is usually limited to a peak deviation of only 1.5–2 kHz to reduce the bandwidth required to about 6 kHz. For broadcasting, a deviation of 75 kHz is widely used and typical receiver bandwidth is usually about 150 kHz.

FM transmissions often use pre-emphasis at the transmitter end and de-emphasis at the receiver end to improve the signal-to-noise performance. In pre-emphasis the higher audio frequencies are boosted and in the receiver the de-emphasis restores the audio signals to give a flat audio response.

For broadcasting wideband, FM is used with a peak deviation of 75 kHz and audio signals up to 35 kHz, giving a bandwidth of 220 kHz. Amateur and CB stations use narrow-band FM with a peak deviation of 5 kHz and voice signals up to 3 kHz, giving a bandwidth of about 16 kHz. Amateur FM channels at VHF are spaced at 25-kHz intervals to reduce adjacent channel interference.

FM is usually achieved in the master oscillator stage of the transmitter by using a VCO circuit where the speech signal is used as the control voltage. This is often achieved by using a variable-capacitance diode connected across the oscillator tuned circuit and applying the modulating voltage to the diode to vary its capacitance and hence alter the oscillator frequency.

Single-sideband mode (A3J)

A single-sideband (SSB) transmission is basically an AM signal from which the carrier component and one sideband have been suppressed before the signal is transmitted. The remaining sideband component still contains the information to be conveyed. This mode of transmission has the advantage that the power that would have been contained in the carrier and the second sideband can now be put into the sideband that is transmitted. For an output amplifier with a given power capability the effective signal is equivalent to an AM signal of about four times the power level.

SSB signals are usually generated by using a balanced modulator to generate an AM signal and remove the carrier component. This double-sideband suppressed carrier signal is then fed through a bandpass filter to remove the unwanted sideband.

Amateur stations transmit the lower sideband (LSB) on the 1.8-, 3.5- and 7-MHz bands and the upper sideband (USB) on other bands up to 28 MHz. On VHF and UHF either sideband may be used, but it seems that most stations use the USB. Aircraft and most utility stations on the HF bands use SSB transmissions.

Bandwidth is equal to the highest modulation frequency and for amateur· telephony this is approximately 3 kHz.

Speech processors

A speech signal consists of quite high-level peaks combined with a relatively low average signal. The maximum modulation is governed by the audio peaks to avoid overmodulation but most of the information is conveyed by the lower level audio components. If the peaks are clipped off, the level of modulation from the rest of the audio signal can be increased to give greater apparent voice power and hence a more readable signal at the receiving end.

Clipping the audio signal causes distortion and introduces high-frequency components which must be filtered out before the signal is applied to the modulator. The audio processing can also be achieved by applying compression to the audio signal. Here the audio gain is varied according to the signal level being input so that a more or less constant level of audio is output whether the speech input is quiet or loud. This has a similar effect to clipping and increases the effective punch of the audio signal when it is received. The amount of processing that can be applied is limited because as the clipping or compression is increased the speech will tend to become less intelligible. In most transceivers the speech processor uses basic analogue circuits but some more modern units are starting to use digital signal processing to improve the audio signal effectiveness.

Carrier power

The carrier power is the amount of power contained in the RF signal from a transmitter when it is generating an unmodulated or FM output signal. Carrier power can be calculated by measuring the RF output voltage across the transmitter load. This assumes that the load is a pure resistance which may be either a dummy load or a correctly matched antenna.

$$\text{Carrier power} = \frac{V_o^2}{R_L} \text{ W}$$

where

V_o = RMS output voltage (V)
R_L = load resistance (Ω)

Peak envelope power (PEP)

The peak envelope power is the power output at the peak of a modulated cycle. This method of measuring power output is used for SSB transmissions where there is no steady carrier power output.

$$\text{PEP} = 0.5 \times \frac{V_p^2}{R_L} \text{ W}$$

where

V_p = output voltage at peak of the modulation cycle (V)
R_L = load resistance (Ω)

This can be measured by using a peak detecting rectifier circuit to derive the voltage for making the measurement. The modulation applied should be a constant-amplitude audio tone when making the measurement. The detector circuit has a load capacitor C and resistor R, and the product RC is much greater than the period of the AF signal used for modulation.

In many cases the PEP measurement can be made using an antenna tuner unit or standing wave ratio (SWR) bridge which is fitted with a peak reading power output meter. Many amateur transceivers have an automatic level control (ALC) system which limits the peak output power to the specified maximum for the transmitter. This effectively applies automatic gain control to the audio to prevent overdriving of the output stages of the transmitter.

Frequency multipliers

A class C RF amplifier runs in a highly non-linear mode and produces a large amount of harmonic frequency output, most of which is filtered out by the output tuned circuit. By tuning the output circuits to twice the input frequency, a frequency doubler can be produced where the output frequency is twice that of the input frequency. Tripler and quadrupler circuits can be produced in a similar way.

For VHF and UHF operation the doubler or tripler action may be produced by simply rectifying the input RF signal to produce a high level of harmonic frequency signals and then selecting the desired harmonic frequency in the output tuned circuit.

Linear amplifiers

The frequency-generation circuits of a transmitter are basically low-power stages. When an SSB signal has been generated, any following power amplifiers needed to produce the desired power output must operate in a linear fashion to avoid distortion of the signal and the generation of unwanted harmonics and other distortion products.

Most amateur radio transceivers generate about 100 W maximum PEP output, so a high-power linear amplifier is often added to increase the power output to the antenna to the permitted limit of 400 W PEP. Linear amplifiers are not essential for a CW or FM signal, although they will generally produce much less trouble with harmonics and interference than a non-linear power stage.

For a linear amplifier using a single valve or transistor, class A or class B operation must be used. In many cases a push–pull circuit using a pair of valves or transistors operating in class B mode may be used to give high efficiency and higher power than a single transistor or valve stage.

Output matching networks

For optimum transfer of power from the transmitter power output stage to the antenna the output impedance of the power amplifier should be equal to the impedance presented by the antenna and ideally both should be pure resistances.

In most cases the output impedance of the amplifier stage will not be equal to that of the antenna, so some form of matching circuit is required to transform the antenna impedance so that it does match that of the amplifier.

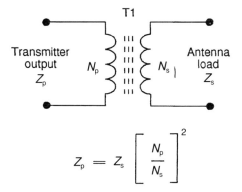

$$Z_p = Z_s \left[\frac{N_p}{N_s} \right]^2$$

Figure 13.6 Output matching using a transformer

Perhaps the simplest approach is to use a transformer for this purpose as shown in Figure 13.6.

The required turns ratio of the transformer can be obtained from the equation:

$$N^2 = \frac{Z_1}{Z_2}$$

where

Z_1 = output impedance of the amplifier
Z_2 = impedance of the antenna

This type of matching transformer would typically be wound onto a ferrite ring core suitable for use at the desired operating frequency. In most cases the output amplifier is matched to a load impedance of 50 Ω and then if required the antenna is adjusted or matched to give an impedance of 50 Ω. The transformer-type matching scheme is often used in amateur transceivers which use transistor output stages.

Alternating matching circuits which may be used to match the output stage impedance to a load impedance of 50 Ω are the PI- and T-type matching circuits shown in Figures 13.7 and 13.8. In both circuits the variable capacitors are adjusted to give the proper impedance match between the transmitter and the load. The PI circuit is usually best for use with valve amplifiers where the amplifier impedance is much higher than 50 Ω, and was widely used in transceivers with valve output amplifiers. The T-type circuit tends to be best for transistor circuits where the amplifier impedance is low. Similar matching networks may be used to match the 50-Ω output impedance of a transceiver to an antenna.

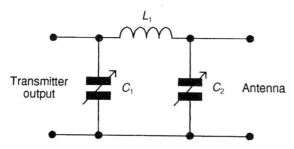

Figure 13.7 PI-type output matching network

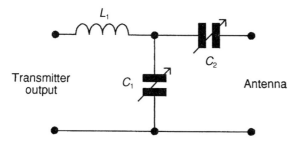

Figure 13.8 T-type output matching network

Commercial transceivers

Most amateurs today tend to use commercially built multi-band transceivers which combine the functions of transmitter and receiver for the HF bands. Often these units will include a general coverage receiver which can tune from 100 kHz to 30 MHz and handle CW, AM and SSB signals. Some transceivers can also handle FM and FSK transmissions. The transmitter section is usually arranged so that it will only operate within the amateur bands. This is controlled by a microprocessor within the unit, and some units intended for commercial use may be programmed to transmit on marine or other bands.

For main station use the transceiver will operate from the mains supply. Typical models are the Yaesu FT1000, Kenwood TS950 and Icom IC737, which provide 100 W output on all bands from 160 to 10 m.

For mobile use there are a number of HF transceivers which are designed to run from a 12-V supply so that they can be connected to a vehicle electrical system. Examples of these are the Yaesu FT747, Kenwood TS450S and Icom IC728. These transceivers can usually deliver 100 W output on all HF bands. Usually there is a matching mains power unit available so that the transceiver can be used as a base station if desired.

Figure 13.9 A typical commercial HF transceiver

For the VHF and UHF bands there is a wide range of handheld transceivers available. Some units can operate on two bands such as 144 and 432 MHz. All current handheld transceivers have facilities for operating duplex through repeaters and usually have a built-in automatic tone burst for accessing the repeaters.

There are also a number of portable units, such as the Yaesu FT290R, which usually have larger battery packs and can therefore operate for longer periods without changing batteries. In addition there are mobile transceivers which are designed to run from a 12-V supply. These normally provide higher power output of around 25 W compared with 3 W for the average handheld or portable unit.

Some amateurs use linear amplifiers to obtain higher output powers up to 400 W on the VHF and UHF bands. These units often contain several amplifier stages and are designed to be driven by a handheld, portable or mobile transceiver with an output of 3–10 W.

14

Receivers

The basic requirements of a radio receiver are that it should be able to select a particular RF signal, amplify it and extract the modulation component to produce either an audible or visual output. Figure 14.1 shows the basic block diagram for a radio receiver for normal broadcast speech or music signals.

The RF signal from the antenna is fed to a high-gain RF amplifier which raises the signal level from a few microvolts to around 0.5 V. This amplifier block also contains selective circuits which pick out the desired signal. The amplifier may operate at the frequency of the incoming signal or, alternatively, most of the signal amplification may be carried out at a lower intermediate frequency.

Once the signal is large enough it is fed to a demodulator which extracts the audio or video modulation component. The type of demodulator used will depend upon the modulation system used in the signal that is being received. After the demodulator a low-pass filter is used to remove any remaining RF signals.

The final block is a further amplifier which drives the output device. In a radio receiver this is an audio power amplifier which drives a loudspeaker.

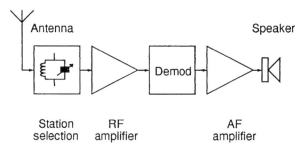

Figure 14.1 Block diagram of basic receiver system

In other applications the signal is amplified and perhaps further decoded to drive a printer, FAX machine, computer or video display.

The crystal set

The simplest form of receiver which may be used for AM broadcast reception is the so-called crystal set, which consists of a tuned circuit and a diode detector as shown in Figure 14.2.

Selectivity is provided by an LC resonant circuit which also provides some voltage amplification of the input from the antenna. L_1 and C_1 effectively form a series resonant circuit and a voltage is induced in series with L_1 by transformer action from the antenna coupling winding on the coil. At resonance the current in the resonant circuit is limited only by the series of resistance of the inductor. The voltage developed across L_1 is given by the current multiplied by the inductive reactance of L_1. This reactance X_L is normally much higher than the resistance, and the ratio between them is known as the Q of the coil. A typical value for Q is 100, so the voltage developed across L_1 would be 100 times the input voltage induced from the antenna winding.

The higher the value of Q, the more selective the circuit becomes, so that it rejects signals on adjacent frequencies more effectively. The bandwidth at which the voltage has dropped by 3 dB is given by:

$$\text{Bandwidth} = \frac{f}{Q} \text{ kHz}$$

where

f = resonant frequency (kHz)
Q = Q factor of the inductor

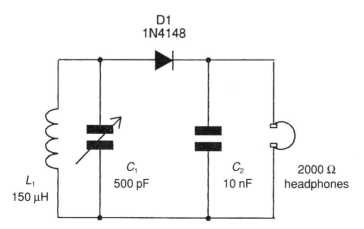

Figure 14.2 A simple crystal set for AM reception

For a circuit with a Q of 100 working at 1000 kHz, the bandwidth would be 10 kHz. The higher the Q, the sharper will be the response of the circuit and the narrower the bandwidth. Practical tuned circuits can provide Q values up to about 200.

The detector circuit is a simple diode rectifier which in a modern-type receiver would use a germanium- or silicon-type semiconductor diode. The rectifier feeds a low-pass filter which removes the RF component from the diode output, leaving the AF component, which then drives the high-impedance (2000-Ω) headphones.

With a good long wire antenna this type of receiver can provide reasonable reception of the stronger medium-wave broadcast signals. The main disadvantage of this type of receiver, apart from its low sensitivity, is that even with a high-Q tuned circuit it will be difficult to separate two strong signals which are on nearby frequencies.

Tuned radio frequency (TRF) receiver

Figure 14.3 shows the block diagram of a simple tuned radio frequency (TRF) or 'straight' receiver. Here signals from the antenna are amplified by one or more RF amplifier stages and then demodulated to give an audio signal. The demodulated audio is then amplified to drive a loudspeaker.

The RF amplifier stages are tuned by variable capacitors which are ganged together to ensure that the tuning of the resonant circuits in all stages remains in step as the frequency is changed. For AM signals the detector is basically a half-wave diode circuit similar to that shown in Figure 14.2. In valve-type receivers the RF amplifier stages normally use pentode valves, which give stable operation when tuned over a wide band of frequencies. The detector stage is often a triode valve, which combines the actions of demodulation and audio amplification.

The valve type of TRF receiver may have positive feedback applied to the triode demodulator stage, which has the effect of increasing the effective Q of the tuned circuit which drives the demodulator. This feedback is called regeneration and is usually controlled by a variable series capacitor feeding back from the anode of the triode to a coupling winding on the grid tuned circuit. As the feedback capacitor value is increased, the effective Q of the

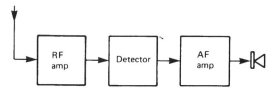

Figure 14.3 Block diagram of a tuned radio frequency (TRF) receiver

grid tuned circuit also increases, giving sharper selectivity and more signal amplification. Eventually the feedback will reach a point where the circuit goes into oscillation. The optimum regeneration setting is just short of the oscillation point when the effective Q of the tuned circuit may be of the order 1000 or more. With regeneration applied, the receiver can provide quite good sensitivity and selectivity on medium- and high-frequency bands.

This type of receiver is not widely used today, since valves are rarely used in receivers and it is difficult to produce stable tuned RF amplifier stages using transistors. Virtually all modern receivers convert the incoming RF signals to a fixed intermediate frequency at which most of the amplification is performed.

Superheterodyne receivers

One problem with TRF receivers is the difficulty of achieving good selectivity and sensitivity whilst maintaining stability of operation. This is due to the requirement for the RF amplifier section to operate over a wide frequency range. The solution is to use a superheterodyne receiver in which the gain and selectivity are carried in a fixed-frequency amplifier.

The block diagram of a superheterodyne receiver is shown in Figure 14.4. Here the incoming signal is mixed with a local oscillator signal to produce a difference frequency component which is called the intermediate frequency (IF). The mixer stage is basically a modulator circuit which produces at its output the sum and difference frequencies of the oscillator and input signals as well as the original signals. The IF is usually the difference between the oscillator and input frequencies but some receivers may use the sum frequency instead. The IF output from the mixer is selected by a bandpass filter which removes the input, oscillator and sum frequency components. Most of the amplification and selectivity is produced in the fixed-frequency IF amplifier. The IF amplifier output is then demodulated to produce audio or video signals which are then further amplified to produce the desired output.

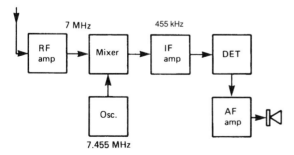

Figure 14.4 Block diagram of a simple superheterodyne receiver

The actual frequency used for the IF amplifier varies according to the application, but the following values are commonly used.

Intermediate frequency	Application
455 kHz	MF and HF broadcast receiver
1.6 MHz	First IF in double superhet
8.8 MHz	Amateur radio transceivers
10.7 MHz	VHF FM receivers
33–39 MHz	UHF TV receivers
35–70 MHz	Amateur radio transceivers
70 MHz	Satellite TV receivers
1000 MHz	Satellite TV receivers

Image interference

In a superheterodyne receiver where the oscillator frequency is above the signal frequency, an input signal with a frequency of $f_o + f_{IF}$ will also produce an output signal at the IF and this will interfere with the wanted signal. This is shown in Figure 14.5. If the oscillator is operating below the signal frequency, the image signal will be at twice the IF below the wanted signal.

This form of interference is called 'second channel' or 'image' interference. In a typical broadcast superhet with an IF of 455 kHz, this form of interference becomes a major problem at signal frequencies above about 10 MHz unless one or more highly selective RF stages are used ahead of the mixer.

Double conversion receiver

The simplest solution to image interference is to use a higher value of IF. Unfortunately, with a higher IF it is more difficult to achieve good selectivity with conventional tuned circuits. This approach is used for TV receivers and FM radio receivers where a wider IF bandwidth is needed.

To overcome the image interference problem in an AM or SSB receiver, a double conversion system may be used as shown in Figure 14.6. Here the

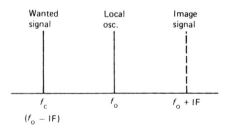

Figure 14.5 Relationship between signal, local oscillator and image frequencies

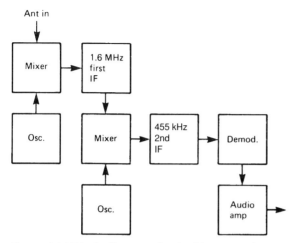

Figure 14.6 Block diagram of a double conversion superheterodyne receiver

first mixer produces an IF of 1.6 MHz, which is called the first IF. By using a higher-frequency first IF, the problem of image interference is eliminated since the image would be 3.2 MHz away from the wanted signal. The first IF signal is then fed to a second mixer which produces a 455-kHz second IF. This lower-frequency 455-kHz IF amplifier can be used to provide most of the required amplification and selectivity.

One problem with superheterodyne receivers operating at higher frequencies, when using a narrow-bandwidth IF, is that the local oscillator frequency will drift with temperature, causing loss of signal and the need for retuning. In some early receivers elaborate precautions were taken to reduce oscillator drift, including placing the oscillator in a temperature-controlled oven.

Most modern receivers use a variation of the double superheterodyne scheme as shown in Figure 14.7. The signal is first up-converted to an IF which is higher than the highest input frequency. This signal is then down-converted to the second IF using a crystal-controlled oscillator. Any drift in the first local oscillator will produce a drift in the first IF frequency, but the down-conversion produces a drift in the second IF which is in the opposite direction so that the two drift effects tend to cancel out.

Most modern amateur radio transceivers use this technique with the first IF in the region 35–65 MHz and the second IF in the region 8–9 MHz. Some receivers use a triple conversion scheme in which the 8-MHz second IF is converted again to a 455-kHz IF for the final amplification and demodulation stages. In many receivers the first oscillator signal is produced by a digital frequency synthesizer, and a digital readout of the frequency is provided rather than the usual calibrated tuning scale. This arrangement has the advantage that the receiver can be set very accurately to any desired frequency.

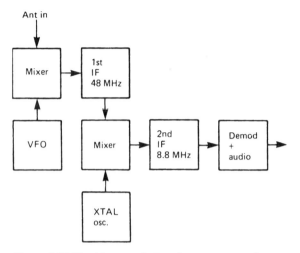

Figure 14.7 Double superheterodyne system using up-and-down frequency conversions for stability

Digital synthesizers

In older receivers the tunable local oscillator uses a standard stable oscillator circuit. With the advent of integrated circuits it is more common to find receivers fitted with digital frequency synthesizers to generate the local oscillator frequency. The synthesizer output is from a VCO. The output of the VCO is also fed to a digital frequency divider chain which has a variable division ratio. The output from the frequency divider is compared with a crystal-controlled reference oscillator and the phase or frequency difference signal is fed back to control the frequency of the VCO. The oscillator frequency will settle down at a frequency of D times the reference frequency, where D is the division ratio of the frequency divider. By varying the division ratio the desired output frequency from the VCO can be set up. A frequency synthesizer of this type provides a very stable oscillator and the frequency can readily be displayed on a digital display.

CW and SSB reception

For the reception of CW and SSB signals a local carrier signal needs to be reinserted at the demodulator stage to render the signals intelligible. For CW the injected signal is set about 800 Hz to 1 kHz away from the frequency of the signal being received, so that the output becomes pulses of audio tone. For SSB reception the local carrier must be injected at the frequency where the suppressed carrier should be, so that the signal can be demodulated to reproduce the original speech signal.

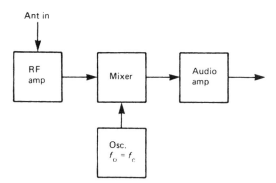

Figure 14.8 Block diagram for a direct conversion receiver

Direct conversion receivers

The block diagram for a direct conversion receiver is shown in Figure 14.8. This receiver is basically a superheterodyne which has an IF of zero. The incoming signal is mixed with a local oscillator of the same frequency, which produces an IF of zero for the carrier component, and the sidebands will produce AF signals corresponding to the modulation signal. A low-pass filter removes any oscillator or input frequency components coming from the mixer output and the resultant audio signal is then amplified to drive a loudspeaker.

In this type of receiver the selectivity is determined by the frequency response of the audio amplifier circuits. Some RF selectivity may be provided by a tuned circuit at the antenna input, since this will help to prevent cross-modulation problems which might be caused by very strong signals on nearby frequencies to the station being received.

For SSB reception the local oscillator need only be within a few cycles of the signal frequency to give acceptable output. This type of receiver is usually called a 'homodyne'. For CW reception the local oscillator is set about 800 Hz away from the incoming CW carrier to provide a suitable AF tone.

For conventional AM reception the local oscillator must be phase locked to the incoming carrier to avoid 'growling' or beat frequency tones. This may be achieved by using a phase-locked loop circuit as the local oscillator. The oscillator and input signal frequencies are compared in a phase detector and the output phase error signal is used to lock the oscillator in phase with the received carrier. This type of receiver with a synchronized local oscillator is called a 'synchrodyne'.

S units

The S unit is a measure of received signal strength. Nominally an S unit is equivalent to a 6-dB increase in signal level at the receiver input. The scale runs from S0 to S9, with the S9 level usually set for an input signal of 50 μV at the antenna input terminals. Actual receiver S meters normally extend beyond the S9 level. Readings above S9 are quoted as S9 + n dB and scales are usually marked up to S9 + 60 dB. The sensitivity of the S meter is often not consistent from one band to another because the basic S meter circuit normally uses the automatic gain control (AGC) voltage as a signal level indication. The reading on the S meter may also vary with adjustment of the RF gain control even when there is no signal present.

Sensitivity

Sensitivity is the ability of the receiver to detect weak signals. This might seem to indicate that a high amplifier gain is required. In fact, the ability to hear weak signals is determined by the amount of noise both from the antenna and from the receiver circuits. Sensitivity is usually quoted for a given ratio of signal to noise. The usual sensitivity figure is based on the number of microvolts of input signal needed to give a 10-dB signal-to-noise ratio for AM and SSB signals. For convenience in measurement, the signal-to-noise ratio is usually defined as the ratio of signal + noise to noise.

A good HF communications receiver will generally have a sensitivity figure of about 0.15 μV for SSB operation with a 3-kHz bandwidth. Since the level of noise is proportional to the bandwidth, a much better sensitivity is possible for CW or RTTY reception, where the bandwidth can be reduced to about 500 Hz. For FM operation on VHF, where the bandwidth is about 15 kHz, a sensitivity figure of about 0.5 μV for 12-dB signal-to-noise might be expected. On frequencies below 3 MHz there is normally quite a lot of static noise present at the antenna input, so there is little point in having a receiver sensitivity much better than 1 μV at these frequencies.

Selectivity

The selectivity of a receiver is its ability to reject signals on frequencies adjacent to the desired signal frequency. The passband is the bandwidth over which the signal voltage falls in amplitude by less than 3 dB. The skirt selectivity is also important, since it defines the rate at which the unwanted signals are attenuated outside the passband. Typically, skirt selectivity is the bandwidth outside which signals are rejected by at least 60 dB. A good response is one where the skirt bandwidth is only a little wider than the passband (Figure 14.9).

Modern communications receivers generally use piezoelectric or crystal filters to define the bandwidth of the IF amplifier. For AM reception, bandwidths of 6 kHz are common, with perhaps a 12-kHz bandwidth for better quality audio. For SSB reception the usual filter gives about 2.4-kHz bandwidth, but some receivers also have a 1.8-kHz filter for SSB reception

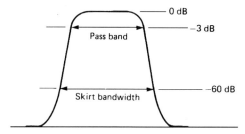

Figure 14.9 Typical frequency response for a crystal-type IF bandpass filter

in heavy QRM. For CW or RTTY a narrow-band filter of 500-Hz bandwidth is common, but again some receivers can be fitted with a narrower 300-Hz filter.

For wideband FM reception the filters will usually have a passband about 150–200 kHz wide. For narrow-band FM as used for amateur radio, the filter passband is usually of the order of 15 kHz.

Communications receivers

Many amateurs today use commercial communications receivers or transceivers in their stations. Most of these modern receivers use a digital synthesizer and digital display to provide an accurate frequency setting and readout in 10-Hz or 100-Hz steps. In this type of receiver the RF amplifier stages are usually untuned, giving a broadband coverage from perhaps 500 kHz to 30 MHz, and a block bandpass filter is included at the input to deal with second channel and breakthrough problems. Most receivers use a bank of filters, with the appropriate filter being switched in for each band. In a general coverage receiver the range 500 kHz to 30 MHz might be divided into perhaps four to six segments, and the corresponding filter is switched in as the tuning frequency is swept through the range.

In this type of receiver the first IF is usually in the region of 40–60 MHz and the second IF is in the region of 8–9 MHz, with selectivity provided by crystal bandpass filters. Mixers are usually of the balanced type using diodes. Some receivers use a third IF of 455 kHz.

Although older communications receivers used individually tuned amplifier stages in the IF circuits, modern designs use the broadband amplifier block technique with the bandpass filters at the amplifier input as shown in Figure 14.10.

Typical commercial HF receivers are the Yaesu FRG100, Kenwood R5000, Icom R72E and JRD NRD535. These all cover the range from 100 kHz to 30 MHz. For VHF and UHF reception the Yaesu FRG9600 and Icom R7100 may be used, but many listeners tend to use scanner receivers on these bands.

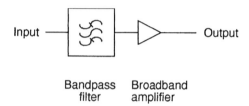

Bandpass   Broadband
filter   amplifier

Figure 14.10 Block filter and broadband amplifier scheme used in many modern receivers

Scanner receivers

With the introduction of digital synthesizers for receiver tuning and broad-band RF amplifier circuits it becomes possible to produce a receiver which can automatically scan through the frequencies in a radio band to find all of the signals that can be heard. The scanner system is usually linked to the AGC circuit and may be arranged to stop its scan when it detects a signal of a suitable level. In the autoscan mode the receiver will stop for a few seconds on each signal and then continue its scan. If a signal sounds inter-esting, the scan can be manually stopped to allow the user to identify the signal.

Modern scanners were originally designed for use at VHF and UHF with either AM (for aircraft bands) or narrow-band FM for amateur and various utility stations. Some receivers also have the option of wideband FM for listening to VHF broadcast stations. Some scanners have extended coverage down to the HF bands but their performance leaves a lot to be desired on these bands. Tuning is usually in 5-kHz steps with a bandwidth of perhaps 15-kHz, which makes these receivers almost useless on the HF amateur bands. Later models have narrow-band filters for receiving SSB and finer tuning steps, which makes them more suitable for use on the HF bands, but their performance is somewhat inferior to that of a dedicated HF receiver.

Most modern scanners include a bank of 200 to perhaps 1000 memories in which the frequency and mode can be stored. Having stored a selection of frequencies it is then possible to scan through them looking for signals. When the scan detects a signal it will stop. The unit may also be programmed so that it will continue scanning if no audio is detected or after some fixed delay time. This prevents the scan being locked up by a random carrier. Often the memories are grouped into banks of perhaps 10 or 20 memories, and individ-ual banks can be scanned rather than all of the memories. Thus one bank might be used for airband stations and another for amateurs.

Some scanners are designed to cover just the aircraft bands or perhaps the amateur and broadcast bands. These usually have two or more switched ranges of operation. More sophisticated scanners provide continuous cover-age from perhaps 25 MHz to around 1500 MHz, whilst some of the newer models include coverage of the medium- and short-wave bands as well.

Figure 14.11 Typical HF communications receiver

Figure 14.12 Typical wideband scanner receiver

Typical examples of base station scanners are the AOR AR3000A and Nevada MS1000. The Yaesu 9600 and Icom 7100 can also be used as base station scanners. There are also many handheld scanners available. Some are simple, such as the Yupiteru VT125 and VT225, which just cover the aircraft bands. Other units, such as the AOR AR2000 and Yupiteru MVT7000, cover HF, VHF and UHF for AM, FM and wide FM modes. Others such as the Yupiteru MVT7100 include facilities for receiving CW and SSB signals over a range from 550 kHz to 1650 MHz.

15

Instruments and measurements

For the radio amateur and electronics experimenter it will be necessary, from time to time, to make use of various instruments in order to measure circuit parameters or to diagnose faults in operation of the station equipment. The instruments required may range from a simple multimeter to more complex units such as oscilloscopes and digital counters or frequency meters. For some measurements more sophisticated laboratory instruments such as logic analysers and spectrum analysers may be needed but these devices are very expensive and unlikely to be available to the average enthusiast, although professional service departments may well have such equipment. In this chapter we shall look at the principles involved in the various instruments and some of the techniques for using them in amateur applications.

Moving coil and moving iron meters

The standard workhorse for most measurements of voltage and current is a multi-range meter, which is usually based on a moving coil meter movement. The basic meter device consists of a coil of wire mounted on a spindle and placed between the poles of a permanent magnet. When a current is passed through the coil it causes a magnetic field to be generated along the axis of the coil. The magnetic field produced by the coil interacts with the magnetic field from the magnet and this causes a mechanical force to be exerted on the coil. Since the coil is fixed to the spindle, the effect is to make the coil and spindle rotate relative to the magnet. Current is fed to the coil through two hairsprings similar to those used in a clock. The springs provide a restraining force for the coil and act as dampers. If a pointer is fixed to the coil, then with a suitably calibrated scale we have the basis of the moving coil meter.

   An alternative type of meter which may be encountered is the moving iron variety. In this type of meter an iron vane moves within a coil of wire. When current is passed through the coil, the magnetic field produced by the coil attracts the iron vane and, by using an appropriate pivoting system and a pointer attached to the iron vane, a pointer movement is produced which corresponds to the current flowing in the coil. The moving iron meter has a simpler construction than the moving coil type and is usually less expensive. The disadvantages of this type of meter are that the scale is non-linear and the meter is relatively insensitive. This type of meter is often used on battery chargers and for mains voltage indicators where it provides a rough indication of the current or voltage.

Analogue voltmeters

The basic moving coil meter movement measures the current flowing through its coil. A typical meter might need a current of 100 μA to produce a full-scale reading on its dial. The coil resistance might be 1000 Ω, so that the total voltage across the meter when it reads full scale would be 0.1 V. For a practical voltmeter it would be useful to have full-scale readings of perhaps 10 V or maybe 100 V. Let us suppose that we want a voltmeter with a full-scale reading of 1 V. To achieve this we need to increase the effective resistance of the meter so that when 1 V is applied, the current flowing through the meter is limited to 100 μA. This can be done by simply connecting a resistance in series with the meter so that the total resistance in the meter circuit is 10 000 Ω.

To produce a full-scale reading of 10 V the series resistance needs to be increased to 100 000 Ω and for 100 V a 1-MΩ resistor is required. If a meter with a full-scale current of 1 mA had been used, the resistor for 1 V would be 1000 Ω and that for 10 V would be 10 000 Ω. The meter sensitivity is sometimes quoted in terms of ohms per volt, based on the amount of series resistance needed to produce a full-scale reading of 1 V. Thus a 1-mA meter has a sensitivity of 1000 Ω/V, whilst a 100-μA meter has a sensitivity of 10 000 Ω/V.

Since the meter coil has some significant resistance, particularly on the more sensitive movements, the series resistor needed to convert the meter into a voltmeter usually turns out to have a non-standard value. Thus a typical 100-μA meter with a coil resistance of 1000 Ω needs a series resistor of 9000 Ω for a 1-V range, 99 000 MΩ for a 10-V range and 999 000 MΩ for a 100-V range. Dealers who supply the basic meter movement can often supply precision resistors to match the meter for a selection of full-scale voltage ranges. Thus a simple multi-range voltmeter can readily be produced using the circuit shown in Figure 15.1. Here the various series resistors are brought out to separate sockets and one of the test leads is plugged into the appropriate socket to select the voltage range.

Most practical meters use a rotary switch to select the series resistor for each range, thus avoiding the need for lots of separate test lead sockets.

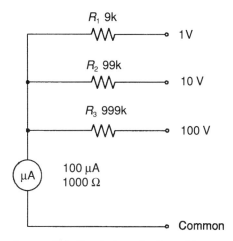

Figure 15.1 Circuit for a basic multi-range voltmeter

Meter protection

With a multi-range voltmeter it is very easy to forget to select the correct range before taking a measurement. If a low voltage range is selected and the actual voltage is much higher than the full-scale value, then excessive current will flow through the meter coil. Most meters are designed to accept a momentary overload of perhaps 8–10 times the rated current but, to prevent damage to the meter movement, some form of overload protection is usually built into the meter circuit. In most meters this simply consists of a pair of diodes connected back to back across the meter coil. The coil voltage is typically around 100 mV at full-scale current and the diodes will limit this voltage to about 300 mV, thus limiting the overload to a reasonably safe level. Some more sophisticated multimeters, such as the AVOmeter, have a magnetic cutout which disconnects the coil if the current becomes excessive and this has to be manually reset in order to restore the meter to its normal working condition.

Voltmeter loading effects

In order to produce a voltage indication, the moving coil meter has to derive its operating current from the circuit on which the measurement is being made. If we are measuring the output voltage from a power supply then the current drawn by the meter is insignificant compared to the output current capability of the power supply, and the voltage reading will be correct.

When the meter is used to measure voltages in a circuit where the normal current levels are relatively small, such as in a transistor amplifier, the

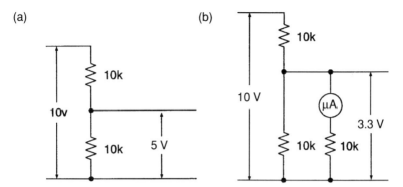

Figure 15.2 Effects of meter loading on a circuit: (a) circuit state without meter connected; (b) circuit state with meter connected

current drawn by the meter can upset the normal working conditions of the circuit and the voltage reading obtained may be wildly inaccurate.

As an example, suppose we have a circuit which consists of two 10-kΩ resistors connected in series across a 10-V supply as shown in Figure 15.2. Since the current through the two main resistors is the same, the voltage at the junction between the resistors must be half the supply voltage, or 5 V. If a 1 kΩ/V meter is used to measure the voltage at the junction of the resistors, it will effectively place another 10-kΩ resistor in parallel with the lower resistor. Thus the effective value of the lower resistor in the chain becomes 5 kΩ. The current through the circuit increases but the voltage at the junction of the resistors with the meter in circuit will be:

$$V = \frac{10 \times 5}{15} = 3.33 \text{ V}$$

which gives a measurement error of some 33%.

If a higher sensitivity meter with an effective resistance of 100 kΩ or maybe 1 MΩ were used the measurement error would be greatly reduced. Thus for measurements of voltages within a circuit it is best to use the highest sensitivity meter available. In service manuals for equipment where typical voltage readings at various points in the circuit are given, the manual will usually specify the sensitivity of the meter used to obtain the readings.

Measuring current

The basic moving coil meter is a current-measuring device but most meter movements used for multi-range test meters have a full-scale current rating of perhaps 100–500 µA which is not particularly useful for practical current measurements. Typically, we would want to measure milliamps or amperes.

If a resistor is connected across the meter coil and this has the same resistance as the coil, then half of the total current flows through the resistor and half through the coil, so that the full-scale current reading is now twice that of the normal meter. The parallel resistor is called a shunt and if it is made, say, one-ninth of the resistance of the meter coil then one-tenth of the total current flows through the meter and the rest through the shunt. In this case the effective full-scale reading of the meter is increased by a factor of 10. By connecting suitable shunt resistors across the meter coil the effective current range can be set to read currents of up to tens of amperes.

In a typical multimeter a number of different shunt resistors may be switched across the meter coil to give current ranges of say 10 mA, 100 mA, 1 A and perhaps 10 A.

Resistance measurement

An analogue meter can be used to measure resistance by applying Ohm's law:

$$I = \frac{V}{R}$$

where V is the voltage across the resistor R, and I is the current flowing through it. If V is constant, then I will be inversely proportional to the resistance R in the circuit.

The basic arrangement using a moving coil meter is shown in Figure 15.3. To set up the meter scale the test leads are shorted together and the

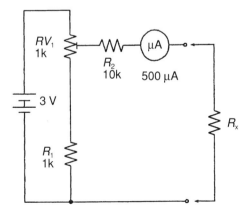

Figure 15.3 Basic circuit for a simple ohmmeter

preset RV_1 is adjusted until the meter reads full scale. This adjustment corrects for changes in battery voltage. The current is mainly controlled by the resistor R. If the test leads are now connected to an external resistor, the total resistance in the circuit is increased and the current reading falls. When the external resistance is equal to the series resistance inside the meter, the current reading will have fallen to half-scale. The resistance scale is non-linear, with the higher resistance readings being compressed into the lower part of the scale. For reasonably accurate resistance measurements the reading should be in the upper half of the scale. By switching in different values of series resistor a selection of resistance ranges can be provided. For low resistance ranges a voltage of 1.5 or 3 V is generally used and is provided by one or two AA-size cells. For higher resistance ranges a voltage of 9 or 15 V may be used in order to obtain sufficient current.

Wheatstone bridge

Although a multimeter can give an approximate value for a resistance, a much more accurate method uses the Wheatstone bridge circuit shown in Figure 15.4. Here R_1 and R_2 are precision resistors which form what are known as the ratio arms of the bridge. R_3 is the resistor being measured and R_4 is an accurately calibrated variable resistor. The meter connected across the centre of the bridge is a centre zero type with as high a sensitivity as possible.

Let us assume that R_1 and R_2 are equal in value so that the voltage at their junction is approximately half the supply voltage. If the value of R_4 is adjusted so that it is exactly equal to R_3, then the bridge will be balanced so that the voltage levels on each side of the meter are the same. Thus there is no voltage difference applied across the meter and no current flows. If R_4 is altered from this balanced state the voltages at each side of the meter become different and current flows through the meter.

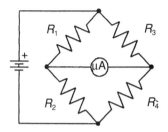

Meter is centre zero type

Figure 15.4 Wheatstone bridge for the measurement of unknown resistors

In the balanced or null condition the voltages at each side of the meter are given by:

$$\frac{VR_1}{R_1 + R_2} = \frac{VR_3}{R_3 + R_4}$$

which can be rearranged to give the equation

$$R_3 = \frac{R_1}{R_2} \times R_4$$

Thus when $R_1 = R_2$, in the null condition R_3 will be equal to the value of R_4.

By altering the values of R_1 and R_2 the effective resistance range of the bridge can be changed. If we make $R_1 = 10R_2$, then R_3 becomes 10 times R_4 when the bridge is balanced. Similarly, if $R_2 = 10R_1$, then R_3 becomes 0.1 times R_4 at balance. Thus the ratio of R_1 to R_2 can be used to set up different ranges for making measurements of the unknown resistor R_3.

AC bridges

Normally a Wheatstone bridge uses a DC supply to energize the bridge and a meter to act as the null indicator. It is, however, equally practical to use an AC signal to energize the bridge and to use a pair of headphones as the null detector, as shown in Figure 15.5. Usually the oscillator circuit used to energize the bridge runs at a frequency of around 1000 Hz, since this is a convenient frequency for producing a clearly audible signal in a pair of headphones. When the bridge is used, the variable resistor in the bridge is adjusted until the tone in the headphones disappears, when the bridge will be balanced.

An AC bridge can be used to measure capacitance by using the circuit shown in Figure 15.6. Here C_1 is the capacitor to be measured and C_2 is a calibrated variable capacitor. There are a number of other AC bridge arrangements which can be used to measure inductance and capacitance.

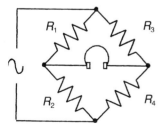

Figure 15.5 Bridge circuit for AC operation

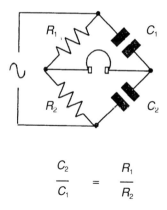

$$\frac{C_2}{C_1} = \frac{R_1}{R_2}$$

Figure 15.6 AC bridge circuit for the measurement of capacitance

One useful type of AC bridge is the noise bridge, in which the signal source is a wideband noise generator. The detector is usually a radio receiver, where the signals from the bridge are coupled to the antenna input of the receiver. Bridge balance is detected when the noise from the receiver falls to a minimum. This type of bridge is suited for the measurement of antenna impedance.

Digital voltmeters

A widely used instrument today is the digital voltmeter, which gives a direct decimal readout of the voltage on a numeric display.

The basic principle of the digital meter is that the voltage input is converted into a numerical count of the time taken to charge or discharge a capacitor.

If a capacitor is charged with a constant current the voltage across it rises linearly with time. If the capacitor voltage is compared with the unknown input voltage we can measure the time it takes for the capacitor to charge from zero to the input voltage. The time measurement is made by starting a counter when the capacitor starts to charge and stopping it when the capacitor voltage equals the input voltage being measured. The counter reading is now directly proportional to the input voltage. If the capacitor value, charging current and clock frequency of the counter are properly chosen, the counter reading can be arranged to correspond to voltage in millivolts. In practice the circuit continually repeats the measurement cycle at perhaps 50 times a second, so that the display reading will also follow slow changes in the voltage being measured.

In practice the simple approach to producing the digital meter just described is difficult to implement with any degree of accuracy, since the reading obtained depends upon the exact value of the capacitor being

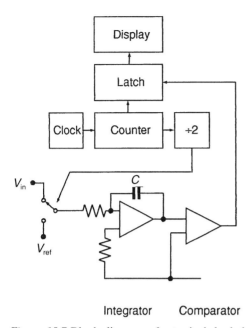

Integrator   Comparator

Figure 15.7 Block diagram of a typical dual slope digital voltmeter

charged, the charging current and the clock frequency of the counter. An alternative approach which overcomes these problems and is used in virtually all modern digital meters is known as the dual slope technique.

The basic arrangement of a dual slope voltmeter is shown in Figure 15.7. In the first part of its operation the capacitor is charged with a constant current which is proportional to the input voltage, whilst the counter counts from zero to its full-scale value. At this point the circuit is switched so that the capacitor is discharged at a constant current derived from an accurate voltage reference, whilst the counter starts to count up from zero again. When the capacitor voltage reaches zero the counter is stopped and its reading is transferred into a storage register which drives the display. The whole cycle then repeats.

Since the capacitor is first charged and then discharged, the value of the capacitor is not critical since any errors during the charge cycle are cancelled during the discharge cycle. Similarly, the clock frequency does not need to be precise as long as the clock runs at a constant speed during the conversion cycle. The accuracy of reading, therefore, is governed mainly by the internal voltage reference.

Most meters have a full-scale reading of 1999, and a decimal point may be set up on the display to give the correct voltage scale, which might range from 1.999 to 1999 V full scale.

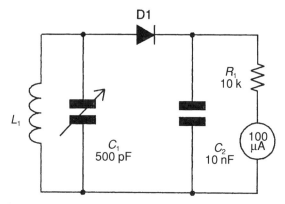

Figure 15.8 Basic absorption wavemeter circuit

A digital meter can be used to measure current by connecting a low-value shunt resistor in series with the circuit carrying the current and then measuring the voltage across the resistor. By suitable choice of shunt resistor the appropriate current scales can be produced on the display.

Absorption wavemeter

One simple method of measuring the frequency of an RF signal is to use an absorption wavemeter. This consists basically of a parallel tuned circuit feeding a diode detector with perhaps a DC amplifier to drive a meter as shown in Figure 15.8. The inductor is mounted outside the unit and when it is brought close to a circuit carrying RF current some of the signal will be coupled into the wavemeter coil. This induced signal will produce an indication on the meter. The tuning capacitor of the wavemeter is then adjusted to obtain a maximum reading on the meter. The capacitor is fitted with a calibrated frequency scale and the frequency of the signal can be read off from the scale.

The absorption meter needs a signal in order to operate, so it is only useful for checking frequencies being generated by oscillators or available at the output of an amplifier. Some care is also needed to ensure that the frequency measured is not a harmonic of the actual frequency present in the circuit. This type of wavemeter is useful for checking that oscillators are running at roughly the desired frequency or checking for harmonic output from a transistor power amplifier. This type of wavemeter can be connected to a short whip antenna and used as a simple field strength meter.

Heterodyne frequency meter

One method of measuring frequency quite accurately is by using a heterodyne wavemeter. This is effectively an accurately calibrated oscillator and a

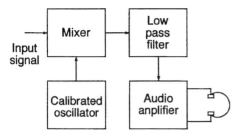

Figure 15.9 Block diagram of a typical heterodyne wavemeter for frequency measurement

mixer circuit as shown in Figure 15.9. The signal to be measured is fed into the mixer circuit, where it is combined with the calibrated oscillator. The output from the mixer consists of the two original signals and also the sum and difference frequencies. A low-pass filter is used to extract the difference frequency and this is fed via an audio amplifier to a pair of headphones. When the oscillator frequency comes within 10 kHz of the input frequency, the difference frequency falls into the audio range and will be heard as a whistle in the headphones. As the frequencies are brought closer together the audio tone drops until it becomes a low-pitched growl. At this point the frequency of the oscillator is within a few cycles of the frequency being measured, and the frequency reading can be obtained from the oscillator dial calibration.

Usually a heterodyne wavemeter will also contain a crystal calibrator which generates harmonics at 1-MHz intervals. This can be used as a reference signal to set up the calibration of the oscillator tuning scale. There may also be a second crystal oscillator generating marker signals at 100-kHz intervals. This type of instrument was widely used by radio amateurs for frequency measurement but today it has been replaced by the digital frequency meter.

Digital frequency meters

Today, frequency measurement is generally carried out by using a digital frequency meter, which gives a direct readout of frequency on a digital display.

The basic circuit arrangement of a digital frequency meter is shown in Figure 15.10. The input signal is amplified and then used as the clock input to a digital decade counter. The counter operation is controlled by a digital timing clock which is derived from an accurate crystal oscillator.

At the start of a measuring cycle the counter is reset to zero and then switched to its counting mode. After a time period of 1 second the counter is stopped and its reading is transferred to a storage register which drives the

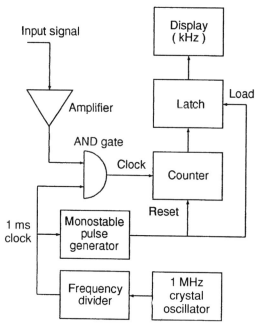

Figure 15.10 Block diagram of digital frequency meter

display. The measurement cycle then starts again. Thus the counter will count the number of cycles of the input frequency that occurred in a period of 1 second and the display will show the frequency in hertz. If the counting period were reduced to 1 ms then the display reading would be in kilohertz.

A frequency meter using conventional logic devices, such as TTL or HCMOS, will usually operate successfully to frequencies of around 30–50 MHz. For higher frequency operation a prescaling frequency divider may be used to bring the counter input frequency down to an acceptable value. These prescaler circuits usually divide by ten and often use emitter coupled logic (ECL) circuits. By using one, or more, prescaler counters, frequencies up to 1000 MHz can be measured.

The oscilloscope

For the examination of waveforms an oscilloscope is an essential piece of equipment. The basic oscilloscope consists of a cathode ray tube, which provides a graphical display of the waveform, a signal amplifier, known as the Y amplifier, and a timebase circuit which generates a linear sawtooth signal. In the cathode ray tube a beam of electrons is produced by an electron gun at one end of the tube and this beam is fired at a phosphor screen at the other end of the tube. The phosphor screen, which is coated on to the

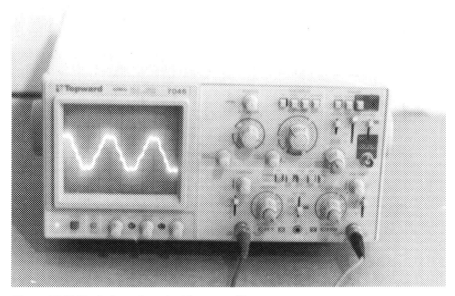

Figure 15.11 Typical modern dual beam oscilloscope

inside of the viewing screen end of the tube, glows when the electrons hit the phosphor. The beam is sharply focused so that a small bright dot appears where the beam hits the screen. Two sets of deflection electrodes in the tube allow the position of the dot to be moved from side to side or up and down when voltages are applied to them.

In the normal operation of the oscilloscope the timebase signal is used to move the spot at constant speed across the screen from left to right. When the beam reaches the right edge of the screen the spot is rapidly shifted back to the left side of the screen and the scan is then repeated. The effect is to produce a horizontal line across the screen which provides a measure of time. The signal to be examined is applied to the Y amplifier and this causes the spot on the screen to be moved in the vertical direction in sympathy with the input signal voltage. Thus if a sinewave signal is applied, the trace on the screen will display a sinewave instead of a horizontal line.

The timebase circuit can be adjusted to give a wide range of sweep rates and would be set to show perhaps one or two cycles of the input signal. The timebase sweep is usually synchronized to the Y input signal so that a stable picture is produced on the screen. Similarly, the gain of the Y amplifier can be varied so that the amplitude of the displayed waveform can be adjusted.

Most modern oscilloscopes have two Y amplifiers, labelled Y1 and Y2, to allow two different signals to be compared or examined at the same time.

The display produces two separate traces on the screen for the two inputs. Usually the two traces are produced by switching the Y1 and Y2 signals alternately to the deflection plates of the tube. Two different modes may be used for this purpose. In the CHOP mode a high-frequency square wave is used to switch the Y1 and Y2 signals to the tube so that each trace is effectively displayed as a series of short dashes. In practice the switching rate is high enough that the traces appear to be solid lines. This mode is generally used for lower frequency signals and has the advantage that the phase relationship between the two inputs is correctly displayed. The other mode is ALTERNATE, where the Y1 signal is displayed on one sweep of the timebase and the Y2 signal on the next sweep. This has the disadvantage that the two traces do not represent signals over the same time period and this must be taken into account when interpreting the display on the screen. This mode is generally used for higher frequency signals, where the chop mode can produce fuzzy displays.

The bandwidth of the Y amplifier determines the highest frequency signals that can be displayed properly. For audio work, a Y amplifier bandwidth of 10 MHz is adequate. For examining video signals or signals in digital logic systems, a bandwidth of 20 MHz or more is desirable.

The oscilloscope screen usually has a graticule which is calibrated with 1-cm squares, and for most oscilloscopes the display area is 10 cm wide by 8 cm high. On the central vertical and horizontal lines the centimetre divisions are subdivided into five segments to allow more accurate measurements of time ($\times$) or amplitude (y) of the displayed waveform. Vertical and horizontal shift controls are provided so that the position of the trace can be moved on the screen to align it with the calibration scales when taking measurements.

Some oscilloscopes provide other alternative display modes. One is ADD, where the signals from the Y1 and Y2 amplifiers are simply added together before being used to produce the trace. This produces a single display trace which is a combination of the two input signals. This is useful for checking phase between the two signals or the location of an event on one signal relative to an event on the other input. The other useful mode is XY, where the Y1 signal is used to drive the Y deflection and the Y2 signal drives the X deflection. This mode is useful for phase measurements between signals of the same frequency or harmonically related frequency. According to the relative phase of the two signals, the trace varies from a diagonal line (in phase) to a circle (90° phase shift).

The oscilloscope will also have a Z input which can be used to modulate the brightness of the trace. This can be useful for putting time markers onto the trace, where they will appear as brighter or darker dots relative to the brightness of the rest of the trace. Normally the oscilloscope will have a Brilliance control for adjusting the brightness of the trace and a Focus control to adjust the sharpness of the displayed trace.

Some oscilloscopes have a dual timebase scheme which allows segments of the horizontal trace to be examined in more detail. Here the A timebase is the normal timebase and the B timebase is a faster scan which can be delayed relative to the start of the A scan. To set this up it is usual to select the AB mode first, which produces a normal display using the A timebase and shows a segment of the display brightened up to indicate where the B timebase is located along the trace. The Delay control can then be adjusted to move the brightened area to the part of the waveform that is to be examined in more detail. When the B mode is selected the small segment of the trace is expanded to fill the width of the screen. By adjusting the Delay control it is possible to move through the waveform and examine it in detail. This mode is useful for examining things like the teletext data on a TV signal.

Some more expensive oscilloscopes provide digital storage facilities. In these instruments the input signals are digitized and stored in a digital memory. The memory is then read out repeatedly to generate the screen display. This type of oscilloscope is particularly useful for looking at transient signals which, once captured in the memory, can be examined at leisure. Another useful feature is that amplitude and time measurements can be made digitally and may be displayed on the screen in numerical form.

Standing wave ratio meters

One useful instrument for the transmitting amateur is the standing wave ratio (SWR) meter. This device can be used to measure the degree of matching between a transmitter and the antenna. When a mismatch occurs on a transmission line, some of the signal is reflected from the load end of the line and this reflected wave interacts with the forward wave down the line to produce a fixed pattern of RF voltage along the line which is known as a standing wave. The SWR is the ratio between the maximum and minimum amplitudes of the RF voltage along the line. When the line is matched, there is no reflected wave and the amplitude of the RF voltage along the line is constant, so the standing wave ratio is 1.0.

The basic SWR meter measures the current flowing in the line, and from this can measure the forward signal being sent down the line or the reflected signal coming back from the antenna. Some meters are combined with the antenna matching unit and may provide readings of the forward and reflected signals in terms of power. Other types with two meters use one meter for forward signal and the other for reflected power or SWR. To read SWR the sensitivity is adjusted so that the forward signal meter reads full scale. SWR meters with a single meter usually have a switch which is first set to calibrate, the meter is adjusted for full scale, and when the switch is set to SWR the meter is calibrated to give a direct readout of the SWR.

Most SWR meters are designed to operate with a 50-Ω coaxial feeder line, and when the antenna impedance is 50 Ω there will be no reflected power and the SWR will be 1.0. If the antenna impedance is not 50 Ω the value for SWR will rise. A modern transmitter with a solid-state output amplifier can usually tolerate SWR values up to about 1.5 or perhaps 2.0, but beyond this level the transistors may be damaged if run at full power. Most commercial HF transceivers have a built-in SWR detector circuit which automatically shuts down the drive to the amplifiers if the SWR is too high. Although the ideal SWR is 1.0, it is probably not worthwhile striving to achieve this if the measured SWR is already 1.2 or so when operating on the HF bands. On the VHF and UHF bands SWR becomes more important, since the losses in coaxial cables are much higher at these frequencies and a mismatched line could cause considerable loss of power if the line is of any great length.

16

Electromagnetic compatibility

One important factor which can affect the operation of amateur radio equipment is electromagnetic compatibility (EMC). This covers all aspects of interference between various pieces of electronic equipment. In the case of an amateur transmitter this might involve harmonic or other radiation from the transmitter which causes interference to broadcast or TV receivers in the immediate area. For amateur receivers the problem will usually manifest itself in the form of interference to reception caused by machinery or other electrical equipment in the local area. With the increasing use of computers in the amateur shack, problems can be encountered with interference from the computer itself which can make reception, particularly of weak signals, very difficult.

Electromagnetic interference can be conveyed between two pieces of equipment either by conduction through wiring which is common to both pieces of equipment or by magnetic or electrical radiation from one piece of equipment to the other. When interference problems occur, the source of the problem and the means of transfer need to be identified. At this stage the cause may be identified as due to improper operation of the equipment which is the source of the interference, which can then be dealt with. Where the source is operating correctly, the addition of filters will often cure the problem. In many cases the problem is caused by poor design of the equipment being interfered with and may only be resolved by modifying that unit.

Conducted interference

Interference conducted via the supply mains can cause problems with amateur reception. This usually takes the form of clicks caused by the operation of thermostats and other switches. A more severe problem is the buzz caused by an electric motor. If the interference is being conducted via the power cables, this problem can be resolved by inserting an RF suppression

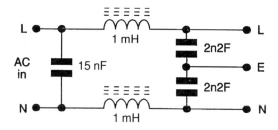

Capacitors rated at 250 V AC

Figure 16.1 Low-pass filter for mains input line

filter into the mains input leads of the receiver. The filter is basically a low-pass LC filter as shown in Figure 16.1. The filter allows the 50- or 60-Hz power to pass through with no attenuation but frequencies in the AF and RF ranges are greatly reduced, thus blocking any interfering signals coming in via the mains supply. Most modern communications receivers have built-in RF suppression filters in the mains circuits.

A radio transmitter may send high-frequency signals back into the power mains via its power supply or by magnetic coupling between the RF circuits and the mains input wiring. Once again, a low-pass filter in the mains input line should prevent the interference signals from being injected back into the mains supply. Modern amateur transmitters usually include these suppression filters as standard. Home-constructed transmitters should include a mains filter to prevent problems. Some home computers do not have mains filters built in and may cause problems if the high-frequency switching pulses from the computer circuits get back into the mains supply. Computers such as the Commodore Amiga and Atari ST and most IBM compatible PCs which have to meet US and German EMC requirements will have built-in mains filters.

Where a computer is connected to a receiver for use with RTTY or SSTV, some problems may be encountered with interference conveyed via the interconnecting wires from the computer to the receiver. This can usually be resolved by including a low-pass filter in the connecting lead which permits the audio signal to pass but suppresses any RF signals.

Radiated interference

The usual problem with radiated interference from an amateur transmitter is that of harmonics of the transmitted frequency which fall within broadcast bands. With HF transmitters the main problem area is likely to be the VHF FM radio band. Commercial transceivers usually have low-pass output filters to reduce harmonic radiation to some 60 dB below the level of the main HF signal. If harmonic radiation is suspected as a problem, the inclusion of a further low-pass filter between the transmitter and the antenna feeder should resolve the problem.

Transmitters operating in the VHF bands may cause harmonic interference to TV signals in the UHF band. Once again the problem can usually be resolved by adding a low-pass filter into the antenna feed line. In a UHF transmitter where the final frequency is achieved by using frequency doubler or trebler stages before the final amplifier, it is possible that some of the lower frequency signals may reach the antenna and these could cause problems with VHF receivers. In this case a high-pass filter may be used to remove these lower frequency components from the antenna feeder.

Local field problems

One major problem which can be encountered when using a transmitter is caused by the high RF field strength produced in the vicinity of the transmitting antenna. This can cause problems in badly designed audio and radio equipment even when the transmitter is operating correctly.

The problem is usually caused by pickup of the RF signals on unscreened leads in the audio or receiver equipment. An unscreened lead of perhaps 30 cm in length could easily pick up a volt or more of signal, and when this reaches a non-linear circuit such as a diode or a transistor the RF signal is rectified and the modulation component is extracted; this is then amplified by the audio circuit of the equipment and produces an unwanted output signal. If the transmitter is operating using FM, the RF signals picked up within the audio equipment will, when rectified, produce a DC component which can upset the operation of the amplifier and cause distortion of the output signals.

With this type of problem one solution at the transmitting end might be to resite the antenna if it is close to the equipment being interfered with. This would apply where the transmitting antenna is located indoors. If this is not practical, then the problem must be resolved in the equipment being interfered with. Pickup of the RF field can be reduced by screening all input leads to the equipment or by inserting low-pass filters in these lines to block any RF signals picked up in the external cables. Another solution is to enclose the equipment in a screened box so that the RF field cannot reach sensitive circuits.

TV interference which is caused by the strong local field may be cured by placing a high-pass filter in the TV antenna lead near the receiver end. This filter will greatly reduce any HF signal which has been picked up on the TV antenna or feeder cable whilst allowing the TV signals to pass through. A similar approach might be used on an FM VHF receiver, where the interference is caused by pickup of the strong HF field rather than by radiation of a harmonic of the operating frequency which falls in the VHF broadcast band.

In the case of an amateur transmitter interfering with a neighbour's audio or radio system where the transmitter is not at fault, the amateur should not get involved with making modifications to the audio or radio equipment itself. Attempts to resolve the problem should be limited to fitting external filters. If this does not resolve the interference, the problem should be referred to the Radio Authority and responsibility for resolving the deficiencies in the audio or radio equipment will be that of the owner or manufacturer rather than the radio amateur. The RSGB can often help in these cases, when their EMC representative for the local area can investigate and suggest solutions which might be tried before the problem is referred to the Radio Authority.

Harmonic radiation

One problem which may be encountered with home-constructed transmitters is that of harmonic interference. Most RF power amplifiers will produce not only the desired output signals but also harmonics of that signal. Whilst these harmonics will be at a lower level than the primary output signal, they can cause severe interference to nearby receivers if they are allowed to reach the transmitting antenna.

To deal with harmonic problems a low-pass filter is placed in the antenna output line. To prevent direct radiation from the transmitter wiring, the unit should be enclosed in a shielded box so that RF signals only exit via the antenna connector. Normally, harmonics should be suppressed to 40 dB below the primary output signal. In some cases where the interference is in a particular band, such as VHF FM radio or VHF TV, a notch or bandstop filter may be used which further suppresses any signals in that band.

Computer interference

One item of equipment that is widely used in amateur radio stations today is the home computer, and this can present problems with interference to radio reception.

A typical home computer contains logic circuits which are switching at high frequencies. Often the primary clock frequencies will be in the order 5–10 MHz and many other frequencies may also be present. Some computers are simply built onto a single circuit board, and the case is made of plastic so that there is no screening to prevent radiation from the circuits on the board. If the computer is close to the receiver it can generate severe interference on most of the HF bands and well into the VHF region. Another source of potential RF interference is the monitor or TV receiver used as a display for the computer. Signals may also be radiated by interconnecting cables between the computer and any external disk drives.

Most modern home computers, such as the Atari, Amiga and PC compatible types, have internal screening or metal cases which will greatly reduce any radiated interference. These computers are usually designed to comply with regulations specifying the level of radiation which is permitted. Despite this, interference may be experienced on some frequency bands.

Most computers are fitted with mains filters, so conducted interference should be low, but if there are not mains filters in the receiver circuit these should be added close to the receiver. Radiated interference may be reduced by ensuring that the earth points of the receiver, transmitter and computer are tied to a good earthing system. It is advisable to locate the computer a few feet away from the receiver to reduce magnetic coupling between them.

17

Callsigns, countries and zones

Most radio stations use a callsign in order to identify themselves. The callsign consists of a prefix which identifies the country and a further group of letters and numbers which identify the individual station. Each country in the world has been allocated one or more blocks of prefix letters. Amateur callsigns consist of the country prefix followed by a number and up to three letters to identify the station. Thus W3ABC would be located in the USA, as indicated by the W3 prefix, and the ABC is the station identifier. In many countries the prefix is used to indicate the region of the country from which the station is operating. Commercial land-based stations use a three-letter callsign which may be followed by up to three figures to indicate an individual transmitter. Ship stations use a four-letter callsign of which the first one or two letters will identify the country. Most passenger aircraft identify by using their flight number, although an alternative is to use the registration letters of the aircraft. Some airlines have their own callsign name. Thus British Airways flights use the callsign 'Speedbird' followed by the flight number. Military aircraft often use callsigns such as 'Hawk 1' which are allocated for each exercise or mission. Some broadcast stations, particularly in the USA, use three- or four-letter calls such as WABC, KNBC and WHO. Usually, stations in the eastern USA use W callsigns and those in the western states use K callsigns.

Callsign prefix allocations

Prefix codes and the countries to which they are allocated are:

Prefix code	Country
A2A–A2Z	Botswana
A3A–A3Z	Tonga
A4A–A4Z	Oman
A5A–A5Z	Bhutan

Prefix code	Country
A6A–A6Z	United Arab Emirates
A7A–A7Z	Qatar
A8A–A8Z	Liberia
A9A–A9Z	Bahrain
AAA–ALZ	USA
AMA–AOZ	Spain
APA–ASZ	Pakistan
ATA–AWZ	India
AXA–AXZ	Australia
AYA–AZZ	Argentina
BAA–BZZ	China
C2A–C2Z	Nauru
C3A–C3Z	Andorra
C4A–C4Z	Cyprus
C5A–C5Z	Gambia
C6A–C6Z	Bahamas
C7A–C7Z	World Meteorological Organization
C8A–C9Z	Mozambique
CAA–CEZ	Chile
CFA–CKZ	Canada
CLA–CMZ	Cuba
CNA–CNZ	Morocco
COA–COZ	Cuba
CPA–CPZ	Bolivia
CQA–CUZ	Portugal
CVA–CXZ	Uruguay
CYA–CZZ	Canada
D2A–D3Z	Angola
D4A–D4Z	Cape Verde Islands
D5A–D5Z	Liberia
D6A–D6Z	Comoros Islands
D7A–D9Z	South Korea
DAA–DRZ	Germany
DSA–DRZ	South Korea
DUA–DZZ	Philippines
E2A–E2Z	Thailand
EAA–EHZ	Spain
EIA–EJZ	Eire
EKA–EKZ	Armenia
ELA–ELZ	Liberia
EMA–EOZ	Ukraine
EPA–EQZ	Iran
ERA–ERZ	Moldova
ESA–ESZ	Estonia
ETA–ETZ	Ethiopia
EUA–EWZ	Belarus
EXA–EXZ	Khirgyzstan
EYA–EYZ	Tadzhikistan
EZA–EZZ	Turkmenya
FAA–FZZ	France
GAA–GZZ	UK
H2A–H2Z	Cyprus

Prefix code	Country
H3A–H3Z	Panama
H4A–H4Z	Solomon Islands
H5A–H5Z	Bophuthatswana
H6A–H7Z	Nicaragua
H8A–H9Z	Panama
HAA–HAZ	Hungary
HBA–HBZ	Switzerland
HCA–HDZ	Ecuador
HEA–HEZ	Switzerland
HFA–HFZ	Poland
HGA–HGZ	Hungary
HHA–HHZ	Haiti
HIA–HIZ	Dominican Republic
HJA–HKZ	Colombia
HLA–HLZ	South Korea
HMA–HMZ	North Korea
HNA–HNZ	Iraq
HOA–HPZ	Panama
HQA–HRZ	Honduras
HSA–HSZ	Thailand
HTA–HTZ	Nicaragua
HUA–HUZ	El Salvador
HVA–HVZ	Vatican
HWA–HYZ	France
HZA–HZZ	Saudi Arabia
IAA–IZZ	Italy
J2A–J2Z	Djibouti
J3A–J3Z	Grenada
J4A–J4Z	Greece
J5A–J5Z	Guinea-Bissau
J6A–J6Z	St Lucia
J7A–J7Z	Dominica
J8A–J8Z	St Vincent and Grenadines
JAA–JSZ	Japan
JTA–JVZ	Mongolia
JWA–JXZ	Norway
JYA–JYZ	Jordan
JZA–JZZ	Indonesia
KAA–KZZ	USA
L2A–L9Z	Argentina
LAA–LNZ	Norway
LOA–LWZ	Argentina
LXA–LXZ	Luxembourg
LYA–LYZ	Lithuania
LZA–LZZ	Bulgaria
MAA–MZZ	UK
NAA–NZZ	USA
QAA–OCZ	Peru
ODA–ODZ	Lebanon
OEA–OEZ	Austria
OFA–OJZ	Finland
OKA–OLZ	Czech Republic

Prefix code	Country
OMA–OMZ	Slovakia
ONA–OTZ	Belgium
OUA–OZZ	Denmark
P2A–P2Z	Papua-New Guinea
P3A–P3Z	Cyprus
P4A–P4Z	Aruba
P5A–P9Z	North Korea
PAA–PIZ	Netherlands
PJA–PJZ	Netherlands Antilles
PKA–POZ	Indonesia
PPA–PYZ	Brazil
PZA–PZZ	Surinam
RAA–RZZ	Russia
S2A–S3Z	Bangladesh
S5A–S5Z	Slovakia
S6A–S6Z	Singapore
S7A–S7Z	Seychelles
S9A–S9Z	Sao Tome-Principe
SAA–SMZ	Sweden
SNA–SRZ	Poland
SSA–SSM	Egypt
SSN–STZ	Sudan
SUA–SUZ	Egypt
SVA–SZZ	Greece
T2A–T2Z	Tuvalu
T3A–T3Z	Kiribati
T4A–T4Z	Cuba
T5A–T5Z	Somalia
T6A–T6Z	Afghanistan
T7A–T7Z	San Marino
T9A–T9Z	Bosnia-Herzegovina
TAA–TCZ	Turkey
TDA–TDZ	Guatemala
TEA–TEZ	Costa Rica
TFA–TFZ	Iceland
TGA–TGZ	Guatemala
THA–THZ	France
TIA–TIZ	Costa Rica
TJA–TJZ	Cameroon
TKA–TKZ	France
TLA–TLZ	Central African Republic
TMA–TMZ	France
TNA–TNZ	Congo Republic
TOA–TQZ	France
TRA–TRZ	Gabon
TSA–TSZ	Tunisia
TTA–TTZ	Chad
TUA–TUZ	Ivory Coast
TVA–TXZ	France
TYA–TYZ	Benin
TZA–TZZ	Mali
UAA–UIZ	Russia
UJA–UMZ	Uzbekistan

Prefix code	Country
UNA–UQZ	Kazakhstan
URA–UZZ	Ukraine
V2A–V2Z	Antigua
V3A–V3Z	Belize
V4A–V4Z	Saint Kitts and Nevis
V6A–V6Z	Micronesia
V7A–V7Z	Marshall Islands
V8A–V8Z	Brunei
VAA–VGZ	Canada
VHA–VNZ	Australia
VOA–VOZ	Canada
VPA–VSZ	UK
VTA–VWZ	India
VXA–VYZ	Canada
VZA–VZZ	Australia
WAA–WZZ	USA
X5A–X5Z	Serbia
XAA–XIZ	Mexico
XJA–XOZ	Canada
XPA–XPZ	Denmark
XQA–XRZ	Chile
XSA–XSZ	China
XTA–XTZ	Bourkina Faso
XUA–XUZ	Campuchea
XVA–XVZ	Vietnam
XWA–XWZ	Laos
XXA–XXZ	Portugal
XYA–XZZ	Myanmar (Burma)
Y2A–Y9Z	Germany
YAA–YAZ	Afghanistan
YBA–YHZ	Indonesia
YIA–YIZ	Iraq
YJA–YJZ	Vanuatu
YKA–YKZ	Syria
YLA–YLZ	Latvia
YMA–YMZ	Turkey
YNA–YNZ	Nicaragua
YOA–YRZ	Rumania
YSA–YSZ	Salvador
YTA–YUZ	Yugoslavia
YVA–YYZ	Venezuela
YZA–YZZ	Yugoslavia
Z2A–Z2Z	Zimbabwe
ZAA–ZAZ	Albania
ZBA–ZJZ	UK
ZKA–ZMZ	New Zealand
ZNA–ZOZ	UK
ZPA–ZPZ	Paraguay
ZQA–ZQZ	UK
ZRA–ZUZ	Republic of South Africa
ZVA–ZZZ	Brazil
2AA–2ZZ	UK

Prefix code	Country
3AA–3AZ	Monoca
3BA–3BZ	Mauritius
3CA–3CZ	Equatorial Guinea
3DA–3DM	Swaziland
3DN–3DZ	Fiji
3EA–3FZ	Panama
3GA–3GZ	Chile
3HA–3UZ	China and Taiwan
3VA–3VZ	Tunisia
3WA–3WZ	Vietnam
3XA–3XZ	Guinea
3YA–3YZ	Norway
3ZA–3ZZ	Poland
4AA–4CZ	Mexico
4DA–4IZ	Philippines
4JA–4KZ	Azerbaijan
4LA–4LZ	Georgia
4MA–4MZ	Venezuela
4NA–4OZ	Yugoslavia
4PA–4SZ	Sri Lanka
4TA–4TZ	Peru
4UA–4UZ	United Nations
4VA–4VZ	Haiti
4WA–4WZ	Yemen
4XA–4XZ	Israel
4YA–4YZ	Civil Aviation (ICAO)
4ZA–4ZZ	Israel
5AA–5AZ	Libya
5BA–5BZ	Cyprus
5CA–5GZ	Morocco
5HA–5IZ	Tanzania
5JA–5KZ	Colombia
5LA–5MZ	Liberia
5NA–5OZ	Nigeria
5PA–5QZ	Denmark
5RA–5SZ	Madagascar
5TA–5TZ	Mauritania
5UA–5UZ	Niger
5VA–5VZ	Togo
5WA–5WZ	Western Samoa
5XA–5XZ	Uganda
5YA–5ZZ	Kenya
6AA–6BZ	Egypt
6CA–6CZ	Syria
6DA–6JZ	Mexico
6KA–6NZ	South Korea
6OA–6OZ	Somalia
6PA–6SZ	Pakistan
6TA–6UZ	Sudan
6VA–6WZ	Senegal
6XA–6XZ	Madagascar
6YA–6YZ	Jamaica
6ZA–6ZZ	Liberia

Prefix code	Country
7AA–7IZ	Indonesia
7JA–7NZ	Japan
7OA–7OZ	Yemen
7PA–7PZ	Lesotho
7QA–7QZ	Malawi
7RA–7RZ	Algeria
7SA–7SZ	Sweden
7TA–7YZ	Algeria
7ZA–7ZZ	Saudi Arabia
8AA–8IZ	Indonesia
8JA–8NZ	Japan
8OA–8OZ	Botswana
8PA–8PZ	Barbados
8QA–8QZ	Maldives
8RA–8RZ	Guyana
8SA–8SZ	Sweden
8TA–8YZ	India
8ZA–8ZZ	Saudi Arabia
9AA–9AZ	Croatia
9BA–9DZ	Iran
9EA–9FZ	Ethiopia
9GA–9GZ	Ghana
9HA–9HZ	Malta
9IA–9JZ	Zambia
9KA–9KZ	Kuwait
9LA–9LZ	Sierra Leone
9MA–9MZ	Malaysia
9NA–9NZ	Nepal
9OA–9TZ	Zaire
9UA–9UZ	Bueundi
9VA–9VZ	Singapore
9WA–9WZ	Malaysia
9XA–9XZ	Ruanda
9YA–9ZZ	Trinidad and Tobago

List of countries

In this list of countries and their associated amateur callsign prefixes, the CQ zone number and the ITU zone number are also indicated.

Country	CQ zone	ITU zone
Afghanistan: YA	21	40
Agalega and St Brandon Islands: 3B6, 3B7	39	53
Aland Islands: OH0	15	18
Alaska: AL7, KL7, NL7, WL7	1	2
Albania: ZA	15	28
Algeria: 7Z	33	37
Amsterdam and St Paul Islands: FT8Z	39	68
Andaman and Nicobar Islands: VU	26	49
Andorra: C3	14	27

Country	CQ zone	ITU zone
Angola: D2	36	52
Anguilla: VP2E	8	11
Antigua: V2A	8	11
Argentina: LU	13	14/16
Armenia: EX	21	29
Ascension Islands: ZD8	36	66
Auckland and Campbell Islands: ZL9	32	60
Australia: VK1-3, VK5	30	59
Australia: VK4	30	55
Australia: VK6	29	58
Australia: VK8	29	55
Austria: OE	15	28
Azerbaijan: 4J	21	29
Azores: CT2, CU	14	36
Bahamas: C6	8	11
Bahrain: A9X	21	39
Baker and Howland Islands: KH1	31	61
Balearic Islands: EA6	14	37
Bangladesh: S2	22	41
Barbados: 8P	8	11
Belarus: EV, EW	16	29
Belgium: ON	14	27
Belize: V3A	7	11
Benin: TY	35	46
Bermuda: VP9	5	11
Bhutan: A51	22	41
Bolivia: CP1, CP8, CP9	10	12
Bolivia: CP2–CP7	10	14
Bosnia-Herzegovina: T9	15	28
Botswana: A22	38	57
Bourkina Faso: XT	35	46
Bouvet Islands: 3Y	38	67
Brazil: PP1–PP5, PT2, PT9, PY1–PY5, PY9	11	15
Brazil: PP6–PP8, PR, PS, PT7	11	13
Brazil: PU, PV, PW, PY6–PY8	11	13
Brunei: VS5	28	54
Bulgaria: LZ	20	28
Burundi: 9U5	36	52
Cameroon: TJ	36	46
Canada: VE1, VE2, VO	5	9
Canada: VE3	4	4
Canada: VE4, VE5	4	3
Canada: VE6	4	2
Canada: VE7	3	2
Canada: VE8, VY1	1–2	2–5
Cape Verde Islands: D4C	35	46
Caroline Islands: KC6	27	64/65
Cayman Islands: ZF	8	11
Central African Republic: TL8	36	47
Chagos Islands: VQ9	39	41
Chatham Islands: ZL7	32	60
Chile: CE1–CE5	12	14
Chile: CE6–CE8	12	16
China: BV	23/24	43–44
Christmas Islands: VK9X	29	54

Country	CQ zone	ITU zone
Clipperton Islands: FO8	7	10
Cocos Islands: TI9	7	11
Colombia: HK	9	12
Comoros Islands: D68	39	53
Congo Republic: TN8	36	52
Cook Island: ZK1	32	62
Corsica: FC	15	28
Costa Rica: TI	7	11
Crete: SV9	20	28
Croatia: 9A	15	28
Crozet Island: FT8W	39	68
Cuba: CM, CO	8	11
Cyprus: 5B4	20	39
Czech Republic: OK	15	28
Denmark: OZ	14	18
Desecheo Islands: KP5	8	11
Djibouti: J28	37	48
Dominica: J73	8	11
Dominican Republic: HI	8	11
Easter Island: CEO	12	63
Ecuador: HC	10	12
Egypt: SU	34	38
Eire: EI, EJ	14	27
El Salvador: YS	7	11
Equatorial Guinea: 3C	36	47
Estonia: ES	15	29
Ethiopia: ET3	37	48
England: G, M, 2E	14	27
Falkland Islands: VP8	13	16
Faroe Islands: OY	14	18
Fiji: 3D2	32	56
Finland: OH	15	18
France: F, TK	14	27
Franz Josef Land: UA1	16	29
French Guiana: FY7	9	12
Gabon: TR8	36	52
Galapagos Islands: HC8	10	12
Gambia: C5A	35	46
Georgia: 4L	21	29
Germany: DA–DL	14	28
Ghana: 9G1	35	46
Gibraltar: ZB2	14	37
Greece: SV	20	28
Greenland: OX	40	75
Grenada: J3	8	11
Guadeloupe: FG7	8	11
Guam: KH2	27	64
Guantanamo Bay: KG4	8	11
Guatemala: TG	7	11
Guernsey: GU, MU, 2U	14	27
Guinea: 3X	35	46
Guinea-Bissau: J5	35	46
Guyana: 8R	9	12
Haiti: HH	8	11
Hawaii: KH6	31	61

Country	CQ zone	ITU zone
Heard Island: VK0	39	68
Honduras: HR	7	11
Hong Kong: VS6	24	44
Hungary: HA, HG	15	28
Iceland: TF	40	17
India: VU	22	41
Indonesia: YB, YC, YD	28	51, 54
Iran: EP	21	40
Iraq: YI	21	39
Isle of Man: GD, MD, 2D	14	27
Israel: 4X, 4Z	20	39
Italy: I	15	28
Ivory Coast: TU2	35	46
Jamaica: 6Y5	8	11
Jan Mayen Island: JX	40	18
Japan: JA–JR	25	45
Jersey: GJ, MJ, 2J	14	27
Johnston Island: KH3	31	61
Jordan: JY	20	39
Kazakhstan: UN, UO, UP, UQ	17	30
Kenya: 5Z4	37	48
Kerguelan Islands: FT8X	39	68
Kermadec Islands	32	60
Kampuchea: XU	26	49
Kirghizstan: EX	17	30
Kiribati: T3	31	65
Korea: D7, HM	25	44
Kure Islands: KH7	31	61
Kuwait: 9K2	21	39
Laos: XW8	26	49
Latvia: YL	15	29
Lebanon: OD5	20	39
Lesotho: 7P8	38	57
Liberia: EL, 5L	35	46
Libya: 5A	34	38
Liechtenstein: HB0	14	28
Lithuania: LY	15	29
Luxembourg: LX	14	27
Macao: CR9	24	44
Madagascar: 5R8	39	53
Madeira: CT3	33	36
Malawi: 7Q7	37	53
Malaysia: 9M	28	54
Maldives: 8Q	22	41
Mali Republic: TZ	35	46
Malpelo Islands: HK0	9	12
Malta: 9H	15	28
Mariana Islands: KH0	27	64
Marshall Islands: KX6, V7	31	65
Martinique: F	8	11
Mauritania: 5T5	35	46
Mauritius: 3B8	38	53
Mexico: XE	6	10
Midway Island: KH4	31	61
Moldova: ER	16	29

Country	CQ zone	ITU zone
Monaco: 3A	15	27
Mongolia: JT	23	33
Montserrat: VP2M	8	11
Morocco: CN	33	37
Mozambique: C9	37	53
Myanmar: XZ	26	49
Namibia: ZS3	38	57
Nauru: C21	31	65
Navassa Islands: KP1	8	11
Nepal: 9N1	22	42
Netherlands: PA, PD, PE, PI	14	27
Netherlands Antilles: PJ2, PJ3	9	12
Netherlands Antilles: PJ4	9	11
Netherlands Antilles: PJ5–PJ8	8	11
New Caledonia: FK8	32	56
New Zealand: ZL	32	60
Nicaragua: HT	7	11
Niger Republic 5U7	35	46
Nigeria: 5N	5	46
Niue: ZK2	32	62
Northern Ireland: GI, 2I	14	27
Norway: LA–LJ	14	18
Oman: A4X	21	39
Pakistan: AP	21	41
Palmyra and Jarvis Islands: KH5	31	61
Panama: HP	7	11
Papua-New Guinea: P29	28	51
Paraguay: ZP	11	14
Peru: OA	10	12
Philippines: DU	27	50
Pitcairn Island: VR6	32	63
Poland: SP	15	28
Portugal: CS, CT	14	37
Puerto Rico: KP4	8	11
Qatar: A71	21	39
Reunion Islands: FR7	39	53
Rodriguez Islands: 3B9	39	53
Romania: YO	20	28
Russia (Asia): UA9, RA9	16–18	20, 21, 30, 31
Russia (Asia): UAO, RAO	18, 19, 22–26	32–34
Russia (Europe): UA, RA	16	19, 20, 29, 30
Russia (Europe): UA2, RA2	15	29
Rwanda: 9X5	36	52
St Helena: ZD7	36	66
St Lucia: J6	8	11
St Martin: FS7	8	11
St Pierre and Miquelon: FP8	5	9
St Vincent: J88	8	11
Samoa: KH8	32	62
Samoa Western: 5W1	32	62
San Marino: 9A1	15	28
Sao Tome and Principe: S92	36	47
Sardinia: ISO	15	28
Saudi Arabia: HZ	21	39
Scotland: GM, 2M, MM	14	27

Country	CQ zone	ITU zone
Senegal: 6W8	35	46
Seychelles: S79	39	53
Sierra Leone: 9L1	35	46
Singapore: 9V1	28	54
Slovakia: OM S5	15	28
Slovenia	15	28
Solomon Islands: H44	28	51
Somalia: T5	37	48
South Africa: ZR, ZS	38	57
Spain: EA, EB, EC	14	37
Sri Lanka: 4S7	22	41
Sudan: ST	34	47, 48
Suriname: PZ	9	12
Svalbard: JW	40	18
Swaziland: 3D6	38	57
Sweden: SK, SL, SM	14	18
Switzerland: HB	14	28
Syria: YK1	20	39
Tadzhikistan: EY	17	30
Taiwan: BV	24	44
Tanzania: 5H3	37	53
Thailand: HS	26	49
Togo Republic: 5V7	35	46
Tokelau Islands: ZK3	31	62
Tonga: A35	32	62
Trinidad and Tobago: 9Y4	9	11
Tristan de Cunha: ZD9	38	66
Tromelin Islands: FR7	39	53
Tunisia: 3V8	33	37
Turkey: TA	20	39
Turkmenya: EZ	17	30
Turks and Caicos Islands: VP5	8	11
Tuvalu: T2	31	65
Uganda: 5X5	37	48
Ukraine: EM, UQ	15	29
United Arab Emirates: A6X	21	39
Uruguay: CX	13	14
USA: AA–AI, AJ, AK, K, N, W	3–5	6–8
Uzbek: UJ	17	30
Vanuatu: YJ	30, 32	56
Vatican: HV	15	28
Venezuela: YV	9	12
Vietnam: XV	26	49
Virgin Islands: KP2	8	11
Virgin Islands: (British): VP2V	8	11
Wake Island: KH9	31	65
Wales: GW, MW, 2W	14	27
Wallis Island: FW8	32	62
Western Samoa: 5W1	32	62
Yeman Arab Republic: 4W1	21	39
Yeman Peoples Republic: 7O	21	39
Yugoslavia: YT, YU, YZ, 4N	15	28
Zaire: 9Q5	36	52
Zambia: 9J2	36	53
Zimbabwe: Z2	38	53

CQ DX zones

For the CQ world-wide Dx contests organized by *CQ magazine* the world is divided up into a set of 40 zones. These zones are also used for a special award known as the Worked All Zones (WAZ) award. The CQ world-wide zones are as follows:

1 Alaska KL7, Canada VY, VE8 (west of 102°W)
2 Canada VE8 (east of 102°W), parts of VE1 and VE2 north of 50N
3 Canada (BC), USA W6, W7 (Ore, Wash, Idaho)
4 Canada VE3, VE4, VE5, VE6, USA WO, W5, W9, W7 (Wyoming, Montana)
5 Canada VE1, VE2, VO, Bermuda FP8, USA W1, W2, W3, W4 (except Tennessee, Kansas, Alaska), USA W8 (except Michigan, Ohio)
6 Mexico XE
7 Belize VP1, Costa Rica TI, Guatemala TG, Honduras HR, Nicaragua HT, Panama HP, Panama (US) KZ5, Salvador YS
8 Bahamas C6, Barbados 8P, Cayman Islands ZF, Cuba CO, Dominica J7, Dominican Republic HI, Grenada J5, Haiti HH, Guadeloupe FG, Jamaica 6Y, Leeward Islands VP2, St Lucia J6, Martinique FM7, Puerto Rico KP4, Turks and Caicos Islands VP5, Netherlands Antilles PJ5–PJ8, Virgin Islands KV4, Windward Islands VP2
9 Colombia HK, French Guiana FY7, Guyana 8R, Suriname PZ, Trinidad and Tobago 9Y, Venezuela YV
10 Bolivia CP, Ecuador HC, Peru OA, Galapagos Islands HC8
11 Brazil PP–PY, Paraguay ZP
12 Chile CE
13 Argentina LU, Falkland Islands VP9, Uruguay CX
14 Andorra C3, Azores CT2, Balearic Islands EA6, Belgium ON, Denmark OZ, Eire EL, Faroes OY, France F, Holland PA, Germany DL, Luxembourg LX, Monaco 3A, Norway LA, Spain EA, Sweden SM, Switzerland HB, UK G
15 Albania ZA, Austria OE, Bosnia-Herzegovina T9, Corsica FC, Czech Republic OK, Estonia ES, Finland OH, Hungary HA, Italy I, Latvia YL, Lithuania LY, Malta 9H, Russia UA2, RA2, San Marino T7, Slovakia OM, S5, Slovenia, Vatican HV, Yugoslavia YU
16 Moldova ER, Ukraine UR–UZ, EM, EO, Belarus EU, EW, Russia UA1, UA3, UA4, UA6, UA9S, UA9W
17 Kazakh UN–UQ, Kirghiz EX, Tadzhik EY, Turkmen EZ, Uzbek UJ, Russia UA9
18 Russia UA9H, UA9O, UA9U, UA9Y, UA9Z, UA0A, UA0B, UA0H, UA0O, UA0S, UA0T, UA0V, UA0W
19 Russia UA0C, UA0D, UA0F, UA0I, UA0J, UA0S, UA0L, UA0O, UA0X, UA0Z

20 Bulgaria LZ, Cyprus 5B, Greece SV, Israel 4X, Jordan JY, Lebanon OD, Romania YO, Syria YK, Turkey TA
21 Afghanistan YA, Armenia EK, Azerbaijan 4J, Bahrain A9, Georgia 4L, Iran EP, Iraq YI, Kuwait 9K, Oman A4, Pakistan AP, Qatar A7, Saudi Arabia HZ, Yemen 4W, Yemen (PDR) 7O, United Emirates A8
22 Bangladesh S2, Bhutan A5, India VU, Maldives 8Q, Nepal 9N, Sri Lanka 4S
23 Mongolia JT, Russia UA0Y, Western China BV
24 Hong Kong VS6, Macao CR9, Taiwan BV, Eastern China
25 Japan JA, Korea
26 Andaman and Nicobar Islands VU, Kampuchea XU, Laos XW, Myanmar XZ, Thailand HS
27 Carolines KC6, Guam KH2, Marianas KH0, Philippines DU and nearby islands
28 Indonesia YC, Papua-New Guinea P2, Sarawak 9M, Malaysia 9M, Singapore 9V, Solomon Islands H4
29 Australia VK6, VK8, VK9X, VK9Y
30 Australia VK1–VK5, VK7, VK9Z, VK0
31 Baker-Howland Islands KH1, Hawaii KH6, Jarvis Islands KH5, Johnston Islands KH3, Kiribati T3, Kure Islands KH7, Marshall Islands KX6, Midway Island KH4, Nauru C2, Tokelau Islands ZM7, Tuvalu T2, Wake Island KW6
32 Fiji 3D, New Caledonia FK8, New Hebrides YJ, New Zealand ZL, Niue ZK2, Pitcairn Island VR6, American Samoa KH8, Tonga A3, Western Samoa 5W, Wallis and Futuna Islands FW8, Norfolk Islands VK9
33 Algeria 7X, Canary Islands EA8, Madeira CT3, Morocco CN, Tunisia 3V, Ceuta-Melilla EA9
34 Egypt SU, Libya 5A, Sudan ST
35 Benin TY, Cape Verde D4, Gambia C5, Ghana 9G, Guinea 3X, Ivory Coast TU, Liberia EL, Mali TZ, Mauritania 5T, Niger 5U, Nigeria 5N, Senegal 6W, Sierra Leone 9L, Togo 5V, Bourkina Faso XT
36 Angola D2, Ascension ZD9, Burundi 9U, Cameroon TJ, Central African Republic TL, Chad TT, Congo TN, Equatorial Guinea 3C, Gabon TR, Rwanda 9X, St Helena ZD7, Sao Tome and Principe S9, Zaire 9Q, Zambia 9J
37 Djibouti J2, Ethiopia ET, Kenya 5Z, Malawi 7Q, Mozambique C9, Somalia 6O, Uganda 5X
38 Botswana A2, Bouvet Islands 3Y, Lesotho 7P, South Africa ZS, Swaziland 3D, Tristan da Cunha ZD9, Zimbabwe ZE
39 Chagos VQ9, Comoros D6, Heard Island VK0, Madagascar 5R, Mauritius 3B, Reunion FR7, Seychelles S7
40 Greenland OX, Iceland TF, Jan Mayen JX, Svalbard JW

ITU zones

An alternative zone scheme is used by the International Telecommunications Union (ITU).

1 Alaska KL7 (west of 141°W)
2 Southeastern Alaska KL7, Canada (south of 80°N, west of 110°W) VY, VE6, VE7, VE8
3 Canada VE4, VE5, VE8
4 Canada VE3, VE8 (70–90°W)
5 Canada VE8, Greenland OX
6 USA (west of 110°W) W6, W7
7 USA (from 90°W and 110°W) W4, W5, W7, W8, W9, W0
8 USA (east of 90°W) W1, W2, W3, W4, W8, W9
9 Canada VE1, VE2, VO (south of 80°N), St Pierre and Miquelon FP8
10 Mexico XE, Clipperton Islands FO8
11 Bahamas C6, Barbados 8P, Belize VP1, Bermuda VP9, Cayman ZF, Costa Rica TI, Cuba CO, Dominica J7, Dominican Republic HI, Grenada J3, Guadeloupe FG7, Guatemala TG, Haiti HH, Honduras HR, Jamaica 6Y, Martinique FM7, Nicaragua HT, Panama Canal KZ6, Panama HP, Puerto Rico KP4, St Lucia J6, Salvador YS, Trinidad and Tobago 9Y, Virgin Islands KV4, Turks and Caicos VP5, Windward and Leeward Islands VP2, Netherlands Antilles PJ4–PJ8
12 Bolivia (CP1, CP8–CP9), Columbia HK, Ecuador HC, French Guiana FY7, Guyana 8R, Peru OA, Suriname PZ, Venezuela YV, Netherlands Antilles PJ2–PJ3
13 Brazil (PP6–PP8, PR, PS, PT7, PU, PV, PW, PY6–PY8, PY0)
14 Bolivia (CP2–CP7), Chile (CE1–CE5), Paraguay ZP, Uruguay CX, Argentina LU (north of 40°S)
15 Brazil (PP1, PP2, PT2, PY1–PY5)
16 Southern Argentina LU, Chile (CE6–CE8), Falklands VP8
17 Iceland TF
18 Denmark OZ, Faroes OY, Finland OH, Jan Mayen JW, Norway LA, Sweden SM, Svalbard JX
19 Russia UA1N, UA1O, UA1Z, UN1
20 Russia UA1P, UA9J, UA9K, UA9X
21 Russia UA9L
22 Russia UA0A, UA0B, UA0H
23 Russia UA0Q
24 Russia UA0I (135–155°E)
25 Russia UA0X, UA0I (155–170°E)
26 Russia UA0I (170°E to 170°W)
27 Andorra C3, Belgium ON, Eire EI, France F, Holland PA, Luxembourg LX, San Marino T7, UK G, GD, GI, GJ, GM, GU, GW
28 Albania ZA, Austria OE, Bosnia-Herzegovina T9, Bulgaria LZ, Corsica

FC, Croatia 9A, Czech Republic OK, Germany DJ, DK, DL, Greece SV, Hungary HA, Italy I, Malta 9H, Poland SP, Romania YO, Slovakia OM, S5, Slovenia, Switzerland HB, Vatican HV, Yugoslavia YU

29 Armenia EK, Azerbaijan 4J, Belarus EU–EW, Estonia ES, Georgia 4L, Latvia YL, Lithuania LY, Moldova ER, Ukraine UJ–UM, UR–UZ, Russia UA1–UA6 (south of 80°N and west of 50°E)

30 Kazakhstan UN–UQ, Turkmenya EZ, Uzbek UJ, Russia UA4H, UA4N, UA4P, UA4W, UA9A, UA9C, UA9F, UA9G, UA9M, UA9Q, UA9S, UA9W

31 Kirghizstan EX, Russia UA9H, UA9O, UA9U, UA9Y, UA9Z

32 Mongolia JT (west of 110°E), Russia UA0O, UA0S, UA0T, UA0W, UA0Y

33 Mongolia JT (east of 110°E), Russia UA0J, UA0U, UA0V

34 Russia UA0C, UA0D, UA0F, UA0L

35 Russia UA0Z

36 Azores CT2, Madeira CT3, Canary Islands EA8

37 Algeria 7X, Gibraltar ZB2, Morocco CN, Spain EA, Portugal CT, Tunisia 3V

38 Egypt SU, Libya 5A

39 Bahrain A9, Cyprus 5B, Iraq YI, Israel 4X, Jordan JY, Kuwait 9K, Lebanon OD, Oman A4, Qatar A7, Syria YK, Saudi Arabia HZ, Turkey TA, United Arab Emirates A6, Yemen 4W, Yemen PDR 7O

40 Afghanistan YA, Iran EP

41 Bangladesh S2, Bhutan A5, Chagos VQ9, India VU, Maldives 8Q, Pakistan AP, Sri Lanka 4S7

42 Nepal 9N

43 China B

44 Korea, Taiwan BV, Hong Kong VS6, Macao CR9

45 Japan JA

46 Benin TY, Bourkina Faso XT, Cameroon T, Cape Verde D4, Gambia C5, Ghana 9G, Guinea 3X, Guinea-Bissau J5, Ivory Coast TU, Liberia EL, Mali TZ, Mauritania 5T, Niger 5U, Nigeria 5N, Senegal 6W, Sierra Leone 9L, Togo 5V

47 Central African Republic TL, Chad TT, Equatorial Guinea 3C, Sao Tome and Principe S9, Sudan ST (west of 30°E)

48 Djibouti J2, Ethiopia ET, Kenya 5Z, Somalia 6O, Uganda 5X, Aldabra Islands VQ9, Sudan ST (east of 30°E)

49 Burma XZ, Laos XW, Thailand HS, Kampuchea XU, Andaman and Nicobar Islands VU

50 Philippines DU

51 Indonesia YC, Papua-New Guinea P29, Solomon Islands H44

52 Angola D2

53 Comoros D6, Madagascar 5R, Malawi 7Q, Mauritius 3B, Mozambique C9, Reunion Islands FR7, Seychelles S7, Tanzania 5H, Zambia 9J,

Zimbabwe ZE
54 Brunei V8, Malaysia 9M2, Sabah 9M6, Sarawak 9M8, Singapore 9V, Christmas Island VK9X
55 Australia VK4, VK8, VK9Z
56 Fiji 3D, New Caledonia FK8, New Hebrides YJ
57 Botswana A2, Lesotho 7P, South Africa ZS, Swaziland 3D
58 Australia VK6
59 Australia VK1, VK2, VK3, VK5, VK7
60 Australia VK9, VK0, New Zealand ZL
61 Baker-Howland Island KH1, Hawaii KH6, Johnston Island KH3, Kure Island KH7, Midway Island KH4, Palmyra Island KH5
62 American Samoa KH8, Cook Island ZK1, Jarvis Island KP6, WH5, Niue ZK2, Tokelau Island ZM7, Tonga A3, West Samoa 5W, Wallis and Futuna Islands FW8
63 Marquesas Island FO8, Pitcairn Island VR6, Easter Island CE0A
64 Guam KH2, Marianas KH0, East Carolines KC6
65 Carolines KC6, Kiribati T3, Wake Island KH9, Nauru C21, Tuvalu T2, Marshall Islands KX6
66 Ascension Island ZD8, Gough Island ZD9, St Helena ZD7
67 Antarctica OR4, Bouvet Island 3Y
68 Heard Island VK0, FB8X and FB8Z
69 Antarctica (40–100°E) VK0, 4K1
70 Antarctica (100–160°E) VK0, 4K1, KC4
71 Antarctica (160°E to 140°W) ZL5
72 Antarctica (80–140°W) KC4
73 Antarctica (20–80°W) VP8, CE9, LU, KC4
74 South pole KC4
75 Greenland OX, Arctic Canada VE8

The remaining ITU zones 76–89 cover sea areas only.

Call areas in the USA

In the USA the prefix number generally indicates the location of the station. Note that the assigned call prefix depends upon the state in which the station was initially licensed. Under new rules, when a station moves to another call area the original callsign may be retained, so that the prefix may not always indicate the current location of a station.

Prefix	Location
W1	Connecticut, Maine, Massachusetts, New Hampshire, Rhode Island, Vermont
W2	New Jersey, New York
W3	Delaware, Maryland, Pennsylvania
W4	Alabama, Florida, Georgia, Kentucky, North Carolina, South Carolina, Tennessee, Virginia

W5	Arkansas, Louisiana, Mississippi, Oklahoma, New Mexico, Texas
W6	California
W7	Arizona, Idaho, Montana, Nevada, Oregon, Utah, Washington, Wyoming
W8	Ohio, West Virginia
W9	Illinois, Indiana, Wisconsin
W0	Colorado, Iowa, Kansas, Minnesota, Missouri, Nebraska, North Dakota, South Dakota

Prefix letter may be W, K, N or AA–AL with the following exceptions:

AL7, KL7, NL7 and WL7 (Alaska)
KP, KV, NP and WP (Puerto Rico, Virgin Islands)
AH, KH, KX, NH and WH (Hawaii and other US islands)

Canadian call areas

Prefix	Location
VE1	New Brunswick, Nova Scotia, Prince Edward Island
VE2	Quebec
VE3	Ontario
VE4	Manitoba
VE5	Saskatchewan
VE6	Alberta
VE7	British Columbia
VE8	North West Territories
VY1	Yukon
VO1	Newfoundland
VO2	Labrador
VX9	Sable Island

Australian call areas

Prefix	Location
VK1	Canberra
VK2	New South Wales
VK3	Victoria
VK4	Queensland
VK5	South Australia
VK6	Western Australia
VK7	Tasmania
VK8	Northern Territory
VK9N	Norfolk Island
VK9X	Christmas Island
VK9Y	Cocos Keeling Island
VK9Z	Willis Island
VK0	Heard Island
VK0	Macquarie Island

18

World time zones

This table gives local times around the world measured relative to GMT for various countries. Where applicable, summer time is included.

Country	Winter	Summer
Afghanistan	+4½	+4½
Alaska	–10	–10
Albania	+1	+1
Algeria	GMT	GMT
Andaman and Nicobar Islands	+5½	+5½
Andorra	+1	+1
Angola	+1	+1
Argentina	–4	–3
Ascension Island	GMT	GMT
Australia:		
VK6	+8	+8
VK5	+9½	+10½
VK8	+9½	+9½
VK1–VK3, VK7	–10	–11
VK4	–10	–10
Austria	+1	+2
Azores	–1	GMT
Bahamas	–5	–4
Bahrain	+4	+4
Bangladesh	+6	+6
Barbados	–4	–4
Belgium	+1	+2
Belize	–6	–6
Benin	+1	+1
Bermuda	–4	–3
Bolivia	–4	–4
Botswana	+2	+2
Bourkina Faso	+2	+3
Brazil:		
East	–3	–3
Central	–4	–4
West	–5	–5

Country	Winter	Summer
Brunei	+8	+8
Bulgaria	+2	+3
Burundi	+2	+2
Cambodia	+7	+7
Cameroon	+1	+1
Canada:		
VO	–3½	–2½
VE1–VE2	–4	–3
VE3	–5	–4
VE4	–6	–5
VE5–VE6	–7	–6
VE7, VY	–8	–7
Cape Verde Islands	–2	–2
Cayman Islands	–5	–5
Central African Republic	+1	+1
Chad	+1	+1
Chagos Island	+5	+5
Chile	–4	–3
China	+8	+8
Christmas Island	+7	+7
Clipperton Island	–7	–7
Colombia	–5	–5
Comoros Island	+3	+3
Congo Republic	+1	+1
Cook Island	–10½	–9½
Costa Rica	–6	–6
Cuba	–5	–4
Cyprus	+2	+3
Czech Republic	+1	+2
Denmark	+1	+2
Djibouti	ı3	ı3
Dominican Republic	–4	–4
Easter Island	–7	–6
Ecuador	–5	–5
Egypt	+2	+3
Eire	GMT	+1
El Salvador	–6	–6
Equatorial Guinea	+1	+1
Ethiopia	+3	+3
Falkland Islands	–4	–3
Faroe Islands	GMT	GMT
Fiji	+12	+12
Finland	+2	+3
France	+1	+2
Franz Josef Land	+5	+5
French Guiana	–3	–3
Gabon	+1	+1
Gambia	GMT	GMT
Germany	+1	+2
Ghana	GMT	GMT
Gibraltar	+1	+1
Greece	+2	+3
Greenland	–3	–3
Guadeloupe	–4	–4
Guam	+10	+10

Country	Winter	Summer
Guatemala	–6	–6
Guinea	GMT	GMT
Guinea-Bissau	–1	–1
Guyana	–3½	–3½
Haiti	–5	–5
Hawaii	–10	–10
Honduras	–5	–6
Hong Kong	+8	+9
Hungary	+1	+2
Iceland	–1	GMT
India	+5½	+5½
Indonesia:		
West	+7	+7
Central	+8	+8
East	+9	+9
Iran	+3½	+4½
Iraq	+3	+3
Israel	+2	+2
Italy	+1	+2
Ivory Coast	GMT	GMT
Jamaica	–5	–4
Jan Mayen Island	–1	–1
Japan	+9	+9
Johnston Island	–10	–10
Jordan	+2	+2
Kenya	+3	+3
Kerguelen Island	+5	+5
Kiribati	+12	+12
Korea	+9	+9
Kuwait	+3	+3
Laos	+7	+7
Lebanon	+2	+2
Leeward Islands	–4	–4
Lesotho	+2	+2
Liberia	GMT	GMT
Libya	+2	+2
Luxembourg	+1	+2
Macao	+8	+8
Madagascar	+3	+3
Madeira	GMT	GMT
Malawi	+2	+2
Malaysia:		
West	+7½	+7½
East	+8	+8
Maldives	+5	+5
Mali Republic	GMT	GMT
Malta	+1	+2
Mariana Islands	+10	+10
Marshall Islands	+12	+12
Mauritania	GMT	GMT
Mauritius	–4	–4
Mexico:		
East	–6	–6
West	–7	–7
Midway Island	–11	–11

Country	Winter	Summer
Monaco	+1	+2
Mongolia	+8	+8
Morocco	GMT	GMT
Mozambique	+2	+2
Myanmar	+6½	+6½
Nauru	+12	+12
Nepal	+5½	+5½
Netherlands (Holland)	+1	+2
Netherlands Antilles	–4	–4
New Caledonia	+11	+11
New Zealand	+12	+13
Nicaragua	–6	–6
Niger Republic	+1	+1
Nigeria	+1	+1
Niue	–11	–11
Norway	+1	+2
Oman	+4	+4
Pakistan	+5	+5
Panama	–5	–5
Papua-New Guinea	+10	+10
Paraguay	–4	–3
Peru	–5	–5
Philippines	+8	+8
Pitcairn Island	–8½	–8½
Poland	+1	+2
Portugal	GMT	+1
Puerto Rico	–4	–4
Qatar	+4	+4
Reunion Island	+4	+4
Romania	+2	+3
Russia:		
Moscow	+3	+4
Sverdlovsk	+5	+6
Tomsk	+7	+8
Irkutsk	+8	+9
Yakutsk	+9	+10
Vladivostok	+10	+11
Magadan	+11	+12
Rwanda	+2	+2
St Helena	GMT	GMT
St Pierre and Miquelon	–4	–4
Samoa	–11	–11
Sao Tome and Principe	GMT	GMT
Saudi Arabia	+3	+3
Senegal	GMT	GMT
Seychelles	+4	+4
Sierra Leone	GMT	GMT
Singapore	+7½	+7½
Solomon Islands	+11	+11
Somali Republic	+3	+3
South Africa	+2	+2
Spain	+1	+2
Sri Lanka	+5½	+5½
Sudan	+2	+2
Suriname	–3½	–3½

Country	Winter	Summer
Swaziland	+2	+2
Sweden	+1	+2
Switzerland	+1	+2
Syria	+2	+2
Taiwan	+8	+8
Tanzania	+3	+3
Thailand	+7	+7
Togo Republic	GMT	GMT
Tokelau Island	–11	–11
Tonga	+13	+13
Trinidad	–4	–4
Tunisia	+1	+1
Turkey	+2	+2
Turks and Caicos Islands	–5	–4
Tuvalu	+12	+12
Uganda	+3	+3
United Arab Emirates	+4	+4
UK	GMT	+1
Uruguay	–3	–3
USA:		
Eastern	–5	–4
Central	–6	–5
Mountain	–7	–6
Pacific	–8	–7
Vanuatu	+11	+11
Venezuela	–4	–4
Vietnam	+7	+7
Virgin Islands	–4	–4
Wake Island	+12	+12
Yemen	+3	+3
Yugoslavia	+1	+1
Zaire:		
East	+2	+2
West	+1	+1
Zambia	+2	+2
Zimbabwe	+2	+2

Time signal stations

A number of stations around the world transmit time signals. Some of these operate 24 hours a day, giving time signals on the hour or at more frequent periods. Others only operate at certain times of the day. Schedules for these time signal stations are given in the *World Radio TV Handbook*.

Country	Station
Argentina	LOL: 5000, 10 000, 15 000 kHz
	LQB9: 8167.5 kHz
	LQC28: 17 551 kHz
Australia	VNG: 5000, 8638, 12 984, 16 000 kHz (24-hour)
Canada	CHU: 3330, 7335, 14 670 kHz (24-hour)
China	BPM: 2500, 5000, 10 000, 15 000 kHz (24-hour)

Czech Republic	OMA: 50 kHz
Ecuador	HD210A: 3810, 5000, 7600 kHz
England	MSF: 60 kHz (24-hour)
Germany	DCF77: 77.5 kHz
Hawaii	WWVH: 2500, 5000, 10 000, 15 000 kHz (24-hour)
India	ATA: 5000, 10 000, 15 000 kHz
Italy	IAM: 5000 kHz
Japan	JJY: 2500, 5000, 8000, 10 000, 15 000 kHz
Korea	HLA: 2500, 5000 kHz
Russia	RID: 5004, 10 004, 15 004 kHz
	RTA: 10 000, 15 000 kHz
	RTZ: 50 kHz
	RWM: 4996, 9996, 14 996 kHz
Taiwan	BSF: 5000, 15 000 kHz
USA	WWV: 5000, 10 000, 15 000, 20 000 kHz (24-hour)
	WWVB: 60 kHz (24-hour)
Uzbekistan	RCH: 2500, 5000, 10 000 kHz
Venezuela	YVTO: 5000 kHz

Standard frequency transmissions

A number of standard frequency transmissions are available from various countries around the world. Many of these stations also provide time signals. Most of these stations operate on frequencies of 5, 10 and 15 MHz but some use other frequencies. The main stations with their operating frequencies are:

Station	Operating frequency (kHz)
ATA, Delhi, India	5000, 10 000, 15 000
BPM, Xi-an, China	5000, 10 000, 15 000
BSF, Taipei, Taiwan	5000, 15 000
HLA, Taejon, Korea	5000
JJY, Tokyo, Japan	2500, 5000, 8000, 10 000, 15 000
LOL, Buenos Aires, Argentina	5000, 10 000, 15 000
RCH, Tashkent, Uzbekistan	2500, 5000, 10 000
RTA, Novosibirsk, Russia	10 000, 15 000
VNG, Llandilo, Australia	5000, 16 000
WWV, Ft Collins, Colorado, USA	2500, 5000, 10 000, 15 000, 20 000
WWVH, Kauai, Hawaii	2500, 5000, 10 000, 15 000
YVTO, Caracas, Venezuela	5000

WWV and WWVH transmissions include the time at every minute and also propagation bulletins at 15 min past each hour. Time is given both in CW and as a voice message.

Most stations identify by callsign in Morse code and time may be given in Morse or by voice.

19

Codes and abbreviations

A number of codes and abbreviations are used by both amateur and professional radio stations. These were mostly introduced when Morse was the
main mode of communication and they were employed in order to save time
in sending a message. The Q codes and many of the amateur abbreviations
are still used today even when communication is by speech and there is virtually no time advantage in using them.

The Q code
The Q code was introduced primarily for telegraphy using Morse, and
consists of a series of three-letter codes which have specific meanings and
enable a relatively long message to be conveyed rapidly. Many of the Q code
groups are used by radio amateurs to save time and may be used for both
telegraphy and telephony contacts. If a query symbol is sent after the code,
it indicates a question, whereas the Q code by itself indicates a reply. An
example is:

QRA? What is your location?
QRA My location is .....

The series of codes from QAA to QIZ and QKA to QOT are used for
messages relating to aircraft operations and are not listed here.

The codes from QJA to QJZ are used for automatic telegraphy operations.

Code	Meaning
QRA	The location of my station is ...
QRB	The distance between our stations is ...
QRE	My estimated time of arrival is ...
QRF	I am returning to ...
QRG	Your exact frequency is ...
QRH	Your frequency varies
QRI	The tone of your signal is ...
	1 Good
	2 Variable
	3 Bad
QRK	Your signal readability is ...
	1 Bad
	2 Poor
	3 Fair
	4 Good
	5 Excellent
QRL	The frequency is in use
QRM	I am being interfered with
QRN	The channel is noisy (static)
QRO	Increase transmitter power
QRP	Reduce transmitter power
QRQ	Send faster
QRS	Send slower
QRT	Stop sending
QRU	I have nothing for you
QRV	I am ready
QRW	Please inform ... that I am calling his station on ... kHz
QRY	Your turn is number ...
QRZ	You are being called by ...
QSA	Your signal strength is ...
	1 Barely audible
	2 Weak
	3 Fairly good
	4 Good
	5 Very good
QSB	Your signals are fading
QSC	I am a cargo vessel
QSD	Your keying is defective
QSI	I have been unable to break in
QSK	I can hear you between my signals. Break in
QSL	I am acknowledging receipt
QSM	Repeat your last message
QSN	I did hear you on ... kHz
QSO	I can communicate with ... direct
QSP	I will relay your message to ...
QSR	Repeat your call on the calling frequency
QSS	I will use the working frequency
QSU	Send or reply on this frequency
QSV	Send a series of V s
QSW	I am going to send on this frequency
QSX	I am listening to ... on ... kHz
QSY	Change frequency to ... kHz
QSZ	Send each word or group twice

The QTA to QUZ codes are intended for search and rescue functions but the following are useful in amateur radio.

Code	Meaning
QTH	My position is ...
QTR	The correct time is ...
QTS	I will send my callsign for tuning or frequency measurement purposes
QTX	I will keep my station open until ...
QUA	Here is news of ...
QUB	Here is the information requested ...
QUM	Normal working may be resumed

Amateur radio abbreviations

Abbreviation	Meaning
73	Best wishes
88	Love and kisses
TNX	Thanks
TKS	Thanks
BCNU	Be seeing you
HPE	Hope
CUAGN	See you again
TX	Transmitter
RX	Receiver
ANT	Antenna
CQ	General call
PSE	Please
BURO	QSL bureau
TCVR	Transceiver
RIG	Transmitter and receiver
NW	Now
HAM	Radio amateur
CONDX	Propagation conditions
OM	Old man
OB	Old boy
YL	Young lady
XYL	Wife
RPT	Report
OT	Old timer
CU	See you
DX	Long distance or difficult contact
WX	Weather
XTAL	Quartz Crystal
MNI	Many
GD	Good
TU	Thank you
ATU	Antenna tuner unit
ABT	About
AGN	Again
ANI	Any
BK	Break in
CUL	See you later

ES	And
FB	Fine business
GB	Goodbye
GM	Good morning
GE	Good evening
GN	Good night
HI	Laughter
HR	Here
NR	Number or near
NW	Now
HW	How
CPY	Copy
R	Roger (all received OK)
SIGS	Signals
U	You
UR	You are
YR	Your
YF	Wife
WKG	Working
VY	Very
WDS	Words
SKED	Schedule

The phonetic alphabet

A number of phonetic alphabets have been used in the past but the commonly used one today, and the one recommended by the licence, is as follows:

Letter	Code word
A	Alpha
B	Bravo
C	Charlie
D	Delta
E	Echo
F	Foxtrot
G	Golf
H	Hotel
I	India
J	Juliet
K	Kilo
L	Lima
M	Mike
N	November
O	Oscar
P	Papa
Q	Quebec
R	Romeo
S	Sierra
T	Tango
U	Uniform
V	Victor
W	Whiskey
X	X-ray
Y	Yankee
Z	Zulu

20

Logs, reports and QSL cards

Amateur radio stations are usually required to maintain a log of all contacts and other operations of the station. In some countries the requirement to keep a log has been relaxed. The following details refer to the log requirements for stations operating in the UK.

The main station log should give the date and the time of starting the first transmission and the time of ending the last transmission for each period of operation of the station. The log must also indicate the band used and the time of changing to a different amateur band if that is done during the operating session. The actual frequency used need not be recorded although many amateurs do include this for their own reference. If the station is operating unattended, using a mode such as packet radio, then the actual frequency of operation must be included in the log.

The mode of transmission and the power level used must also be recorded. Here again, the times when there is a change of mode or a change of power level should also be recorded. Mode may be recorded by using either the emission symbols, such as A3J, or abbreviations such as CW, SSB and RTTY.

The times of starting and ending all contacts with other stations and their callsigns must be recorded. In addition, the times of all CQ calls must be included even if they were not answered.

The log is usually kept in a book, which must not be of the loose-leaf type and should be filled in after each operating session. Specially printed log books with ruled columns for all of the required log details are available from organizations such as the RSGB and ARRL, but any book may be used provided that the required information is recorded. An alternative method of keeping a log is to use a computer and store the log as files on a floppy disk or tape. It is important that this disk should only contain the log data. This method is convenient when using data transmission modes, such as packet radio, where the computer controlling operation of the station can be programmed to compile the log records automatically.

Most amateurs include the signal reports received from the other stations and the reports given to those stations. The name of the operator and location of the station being worked are also noted. These items and any other notes included in the log are not required by the licence conditions but are useful for reference by the operator.

Short-wave listeners should also keep a log of stations received, with times, frequencies and details of the types of transmission. For broadcast listeners, details of the programme heard may also be included.

Signal reports

Amateur stations normally exchange signal reports which indicate the reception conditions at each end of the contact. These reports normally assess the readability of the signal, its strength and quality. Other information regarding fading and interference may also be exchanged. For the short-wave listener a similar report on reception conditions may be sent to the station received. A number of different report coding systems have been used by amateurs, professionals and listeners to describe the receiving conditions.

The Q code

One system of reporting which has been widely used in the past is based on the Q code which is used by maritime and other professional telegraphy stations. This provides reports on readability, strength and tone.

Signal readability (QRK)

Code	Meaning
QRK1	Bad
QRK2	Poor
QRK3	Fair
QRK4	Good
QRK5	Excellent

Signal strength (QSA)

Code	Meaning
QSA1	Barely audible
QSA2	Weak
QSA3	Fairly good
QSA4	Good
QSA5	Very good

Signal tone (QRI)

Code	Meaning
QRI1	Good
QRI2	Variable
QRI3	Bad

Telephony stations use only the first two codes, so that a signal of 5 and 5 would indicate excellent readability and strength.

The RST code
This code is generally used by amateur stations for reporting on the reception conditions of the received signal. The code consists of three digits representing readability (R), signal strength (S) and tone. For voice transmissions only the R and S part of the code is used. The code meanings are as follows.

Readability

Code	Meaning
R1	Unreadable
R2	Barely readable
R3	Readable with difficulty[p
R4	Good readability
R5	Perfectly readable

Signal strength

Code	Meaning
S1	Barely audible
S2	Very weak
S3	Weak
S4	Fair
S5	Fairly good
S6	Good
S7	Moderately strong
S8	Strong
S9	Extremely strong

Tone

Code	Meaning
T1	Extremely rough hissing note
T2	Very rough AC note, not musical
T3	Rough low-pitched AC note
T4	Rather rough AC note, musical
T5	Musically modulated tone
T6	Modulated tone, slight whistle
T7	Near DC tone, smooth ripple
T8	Good DC tone
T9	Pure DC tone

The tone report is normally used only for CW or Morse signals. Thus a perfectly readable and moderately strong signal with a clean note would be reported as RST579.

For telephony contacts only the R and S parts of the code are used, so a typical report might be R5, S8 indicating a perfectly radable and strong signal.

For some reports, signal strengths of S9 plus a number of decibels may be given. These are based on readings from typical receiver S meters, which are usually calibrated up to S9 + 60 dB. An S point in the range 1–9 is usually taken as being approximately 6 dB. The S9 signal level is usually set for an input signal of some 50 mV at the antenna input of the receiver.

Amateurs may also include comments on interference (QRM) and fading (QSB) after the basic signal report.

The SINPO code

A more comprehensive reporting code is the SINPO code, which gives readings for S (signal strength), I (interference), N (noise–static), P (propagation disturbance–fading) and O (overall reception quality). The ratings are as follows:

Strength (S)

Code	Meaning
1	Barely audible
2	Poor
3	Fair
4	Good
5	Excellent

Interference (I)

Code	Meaning
1	Extreme
2	Severe
3	Moderate
4	Slight
5	None

Noise (N)

Code	Meaning
1	Extreme
2	Severe
3	Moderate
4	Slight
5	None

Propagation disturbance (fading) (P)

Code	Meaning
1	Extreme
2	Severe
3	Moderate
4	Slight
5	None

Overall rating (O)

Code	Meaning
1	Unusable
2	Poor
3	Fair
4	Good
5	Excellent

The report consists of the word SINPO followed by a string of five numbers. If any of the parameters is not reported, its number if substituted by the letter X. This type of report code is often used by short-wave listeners when sending a report to a broadcast station, since it provides more information than the simple signal strength report.

The SINPFEMO code

The SINPFEMO reporting code is similar to the SINPO code but provides more information, since it includes additional reports on F (frequency of fading), E (modulating quality) and M (modulation depth). This code is also useful for making listener reports to broadcast stations.

The SINPO parts of the code are identical to those for a SINPO report and the additional parameters are as follows.

Frequency of fading (F)

Code	Meaning
1	None
2	Slow
3	Moderate
4	Fast
5	Very fast

Modulation quality (E)

Code	Meaning
1	Very poor
2	Poor
3	Fair
4	Good
5	Excellent

Modulation depth (M)

Code	Meaning
1	Continuously overmodulated
2	Poor or no modulation
3	Fair
4	Good
5	Excellent

Here the code is sent as for SINPO by sending the letters SINPFEMO followed by eight numbers with an X inserted for any parameter not reported.

Listener reports

Short-wave listeners regularly send reception reports to amateur, broadcast and utility stations to indicate how those stations are being received at the listener's location.

Many amateur stations will acknowledge listener reports by sending their QSL card. When reporting to an amateur station it is important to remember that if the station that the amateur is working is in the same area as the listening station then the amateur already has a report of how he or she is being received and may not be interested in another report from the same area. It will be more interesting to the amateur if he or she is being received in an area from which he or she cannot hear or contact other stations.

When reporting to a utility station it is best to give a report which covers a period of time, or better a report on reception over perhaps three or four days. Either the RST or SINPO codes can be used for the report. If the station operates on several frequencies, a report on reception on the various frequencies may also be of interest. Many utility stations will verify reports by QSL card or by letter. Reports should normally be sent to the chief engineer.

A report to a broadcast station should cover a period of reception of at least 30 min if possible. Apart from the actual signal report, some notes on the programme content should be included. Names of speakers, topics discussed or titles of music played should be indicated if possible. Broadcast stations usually keep a log of the pieces of music played for copyright purposes. Comments on the programme content will also be welcome. Again, if the station is using several frequencies a report on reception on the various frequencies may be of interest.

QSL cards

Most amateur stations exchange QSL cards to confirm their contact. The card is usually of standard postcard size and carries the callsign, location and

Figure 20.1 Some typical QSL cards received from HF and MF broadcast stations

Figure 20.2 Some typical amateur station QSL cards

details of the station and its operator. Spaces are left for the report, using the RST code, frequency, time and any other comments regarding the contact. Many amateurs display the QSL cards received on the walls of the radio shack. QSL cards may also be needed as proof of contacts made when claiming some of the operating awards that are available. The cards are normally returned with the certificate or trophy associated with the award.

Short-wave listeners can also collect QSL cards by sending reception reports to the stations they have received. Apart from QSLs from amateur stations, most broadcast stations and many of the utility stations will confirm reception either by QSL card or letter.

With the increasing cost of postage it is a good idea if the listener encloses one or more International Reply Coupons (IRCs) with the report to cover the cost of return postage. Broadcast stations will often send a pack of literature and schedule of transmissions with their confirmation of reception. Some South American stations send rather attractive pennants.

QSL bureaux

It is possible to send a QSL card direct to the amateur who has been contacted, but for an active amateur station this could involve sending tens or even hundreds of cards each month, and the postage costs start to become rather high. To help with the handling of QSL cards between amateur stations, most national radio clubs operate a QSL bureau. The amateur operator then sends a whole batch of cards off to his or her national bureau. At the bureau, cards from hundreds of amateurs are then sorted according to the country they are addressed to and a packet of cards is then sent to the bureau in each of those countries. At the receiving end the cards are sorted and sent on in packets to the amateur stations in that country. Usually the bureau waits until it has perhaps six or more cards before it sends them on to the individual amateur.

Cards or reports from listeners may also be sent via the national bureau for the amateur station that has been received. Many QSL bureaux will only handle outgoing cards from their own members, although they will pass on incoming cards for any amateur in that country provided that he has placed a stamped, self-addressed envelope (SAE) in the bureau.

In Britain the RSGB operates the main national QSL bureau and accepts cards for all amateurs in the UK. Outgoing cards to foreign countries are only handled for members of the RSGB. For incoming cards, several submanagers each handle the cards for a group of UK callsigns. Any amateur wishing to receive the cards destined for him or her should send two or three large SAEs to the appropriate bureau submanager, and when enough cards to fill an envelope have accumulated they will be sent on. Members can send batches of cards for all countries to the RSGB bureau and they will then be forwarded on to the appropriate foreign bureaux.

If you wish to send a card direct to another amateur, the address may be found in one of the call books that are published each year. These are directories which list in callsign order the addresses of amateur radio stations. The RSGB *Call Book* lists amateurs in the UK, although some stations withhold their actual address and merely give a town as the location. From America there are two call books, one covering all USA stations and the other covering the rest of the world. These are published annually and are available through the book service of the RSGB and the book services of *Practical Wireless* and the *Short Wave Magazine*. For addresses see Chapter 21.

Addresses of broadcast stations can be found in *World Radio and TV Handbook*, which is published annually. For HF utility stations, many addresses are given in *List of Utility Stations*, which is published annually by J. Klingenfuss in Germany. This book is also available via RSGB and some radio magazines.

21

News bulletins and magazines

Amateur radio news bulletins

Some of the national amateur radio organizations broadcast news bulletins at regular intervals to keep their members up to date on developments in amateur radio. These bulletins will give details of contests, amateur band changes, club meetings, Dx expeditions, satellite orbit parameters, propagation conditions and other items of interest to radio amateurs.

RSGB news bulletin

This is transmitted each week on Sunday at the following times:

3650 kHz (SSB)	0900, 0930, 1000, 1030, 1100, 1800
3640 and 3660 kHz	1130
7047 kHz (AM)	0900, 1100
144.250 MHz (SSB)	0930, 1000, 1030, 1100, 1130
144.525 MHz (FM)	0930, 1000, 1030, 1100, 1130

The RSGB news is also relayed through some 432-MHz repeaters and has been transmitted via amateur satellites during passes over the UK. Most packet BBS stations will also carry a transcript of the current RSGB news bulletin.

ARRL news bulletins

The ARRL transmits daily news bulletins from station W1AW at Newington, Conn, USA. These include propagation forecasts and are transmitted using CW, SSB and RTTY modes. Special emergency bulletins may be transmitted on the hour for SSB, at 15 min past the hour for RTTY, and on the half-hour for CW using the same frequencies as the regular bulletins.

ARRL Morse news bulletins (18 words per minute)
Frequencies: 3580, 7080, 14 070, 21 080, 28 080 kHz.
Daily at 0100, 0400 and 2200 GMT.
Monday to Friday also at 1500 GMT.

During the summer months, when local daylight time is in operation, the transmissions are one hour earlier.

ARRL news bulletins using SSB
Frequencies: 3990, 7290, 14 290, 21 390, 28 590, 50 190 kHz
Daily at 0230 and 0530 GMT.
In summer these times are one hour earlier.

ARRL RTTY news bulletins
The RTTY bulletin is sent first at 45 baud using Baudot code then repeated using ASCII code at 110 baud, and finally transmitted at 100 baud using FEC (mode B AMTOR).

Frequencies: 3625, 7095, 14 095, 21 095 and 28 095 kHz.
Daily at 0200, 0500 and 2300 GMT.
Monday to Friday also at 1600 GMT.

In the summer these times are one hour earlier.

PI4AA news bulletins
The Dutch amateur radio organization VERON provides weekly bulletins of Dx news. These bulletins are broadcast on Fridays at the following times:

Telephony (SSB)
1830 GMT Amateur radio news in Dutch
1845 GMT Dx news in English
2030 GMT News in Dutch
2045 GMT News in English

RTTY Bulletin
This is transmitted first in Baudot (50 baud) then repeated in AMTOR mode B (FEC).

2000 GMT Dx news in Dutch
2015 GMT Dx news in English

Frequencies used are:

3602 kHz
4103 kHz
144.80 MHz (FM)
433.45 MHz (FM)

IARN news bulletins

The International Amateur Radio Network (IARN) broadcasts a bulletin several times each day. The bulletin is usually updated once a week on Saturday. Each broadcast lasts for roughly half an hour. These transmissions are made by K1MAN Belgrade Lakes Maine using SSB. Bulletins often include features on various aspects of amateur radio as well as basic news. Unlike other amateur news transmissions, the IARN news is a bit like a normal broadcast with features, news and reports from correspondents.

Transcripts of these news transmissions are also available on some packet BBS stations in the USA.

Frequency: 14 275 kHz
Times: 2200 GMT

Radio magazines

There are a number of magazines available which are of interest to the radio amateur and short-wave listener. These usually contain amateur radio news, reports on band activity and technical articles on radio topics.

Radio Communication, monthly
RSGB, Cranborne Rd, Potters Bar, Herts.
This is the official magazine for members of the RSGB.

Ham Radio Today, monthly
Argus Specialist Publications, Argus House, Boundary Way, Hemel Hempstead, Herts. HP2 7ST.
Magazine for radio amateurs with news, equipment reviews, technical articles and construction projects.

Practical Wireless, monthly
Arrowsmith Court, Station Approach, Broadstone, Dorset BH18 8PW.
Magazine for radio amateurs with news, construction projects, technical articles with columns on band activity, satellites, packet and propagation.

Short Wave Magazine, monthly
Arrowsmith Court, Station Approach, Broadstone, Dorset BH18 8PW.
Magazine mainly for listeners with equipment reviews, band activity reports, technical articles and columns covering airband, marine, satellites, scanners and broadcast bands.

CQ-TV, monthly
BATC Publications, 14 Lilac Ave., Leicester LE5 1FN.
Journal of the British Amateur Television Club with articles and information on FS TV and SS TV.

QST, monthly
ARRL, 225 Main St, Newington, Conn., USA.
Official journal of the American Radio Relay League. Includes news, construction projects and technical articles.

73 Magazine, monthly
73 Magazine, Pine St, Peterborough NH, USA.
Independent magazine with technical articles and construction projects for radio amateurs.

Citizen's Band, monthly
Argus Specialist Publications, Argus House, Boundary Way, Hemel Hempstead, Herts. HP2 7ST.
Magazine for users of CB radio in the UK. Includes articles on CB activities and some construction projects.

Useful books
There are a number of useful books available for the radio amateur and short-wave or VHF listener. Some of these provide up-to-date information on frequencies and schedules of broadcast or utility stations of various types.

Guide to Utility Stations
Klingenfuss Publications
Published annually, this lists all short-wave utility stations that have been active in the previous year, giving modes of operation and times heard. The book also includes schedules and frequencies of press and weather RTTY stations and of weather FAX stations.

Guide to Facsimile Stations
Klingenfuss Publications
Published annually, this lists frequencies and schedules for FAX stations operating on the short-wave bands and includes details of weather satellite FAX transmissions. Many examples of FAX charts and pictures are included.

World Radio TV Handbook
Billboard Publications
This annual book is the most comprehensive guide to radio and TV stations with their frequencies, schedules and addresses. Also includes articles on propagation conditions and receiver reviews.

Amateur Radio Call Book
Radio Amateur Call Book Inc., 925 Sherwood Dr, Lake Bluff, Illinois USA.
This annual publication lists callsigns, names and addresses of radio amateurs. It is published in two volumes, one for USA callsigns and the other for callsigns of stations outside the USA.

RSGB Call Book
RSGB Publications
This annual book lists callsigns, names and locations of radio amateurs in the UK and Eire. The book also includes much useful information on repeaters, beacons and band plans.

Contests and awards

To encourage amateur station operation, a number of contests are held throughout each year in which the aim is to score points by working as many stations as possible in a specified time period. There is also a wide range of operating awards available to radio amateurs. The awards may be attained by working a specified number of stations in one or more countries or continents. In most cases an attractive scroll is received as confirmation of attaining the award and these scrolls are usually displayed on the wall of the station shack.

Contests

Most contests take place at the weekend over a period of either one or two days. During the contest the station may operate on a single band or several bands, depending upon the contest rules. Most stations have a single operator throughout the contest period, but in some contests club stations may enter with several operators manning the station and may have two or more stations operating simultaneously on different bands.

A log is kept of every station worked during the contest. Usually the stations will exchange a serial number as part of the signal report, which helps when the logs are checked for allocation of points. Thus a typical report might be 59075, which indicates that this is the 75th station worked so far. In some contests a zone number or US state abbreviation may be used instead of a serial number in the signal report. After the contest a copy of the contest log is sent to the body organizing the contest and acts as the station entry for allocation of points and placing in the contest results.

One form of contest is the field day, in which stations entering the contest set up portable stations. Some contests operate on a cumulative basis over a period of perhaps a month.

Some of the major HF contests throughout the year are:

Contest	Month
CQ Worldwide Dx 160-m CW contest	January
ARRL International Dx CW contest	February
REF Frenmch Dx Phone/CW contest	February
Dutch Dx Phone/CW contest	February
CQ Worldwide 160-m Phone contest	February
ARRL International Dx Phone contest	March
CQ WPX (worked prefixes) SSB contest	March
RSGB British Commonwealth CW contest	March
RTTY contest	March
CQ WPX CW contest	May
Russian 'M' Phone/CW contest	May
All Asian Dx Phone contest	June
ARRL Field Day	June
RSGB Field Day	June
IARU HF Championship	July
All Asian Dx CW contest	August
European Dx CW contest	August
European Dx Phone contest	September
CQ Worldwide Dx Phone contest	October
CQ Worldwide Dx CW contest	November
Canada Phone/CW contest	December

National societies generally run various contests for VHF and UHF activity which often operate on a cumulative basis where points are scored in several sessions over a period of time.

The Maidenhead Worldwide QRA Locator System

In recent years a new QRA locator system has been introduced which provides location codes for world-wide use. This is generally known as the Maidenhead system and has replaced the earlier European system.

The entire world surface is divided up into a matrix of large squares, each of which is 20 degrees of longitude by 10 degrees of latitude. These squares are identified by a two-letter code, with the first letter indicating longitude position and the second representing latitude position. The longitude letter starts at 180°W with letter A and ends with letter R, which extends to 180°E. For north–south position the squares start with A at the south pole and progress northwards to letter R which extends to the north pole.

Each major square is divided into 100 smaller squares arranged as a 10 by 10 matrix. Here a two-digit number code is used, with the first digit giving the east–west position and the second digit giving the north–south position. These squares are 2 degrees of longitude wide and 1 degree of latitude high.

Each of these secondary squares is divided into a final matrix of 24 by 24 small squares. Each of these squares is defined by a final two-letter code using letters from A to X. These final squares represent an area of 5 minutes of longitude by 2.5 minutes of latitude.

Operating awards

There are many operating awards which can be gained by radio amateurs. The main awards are organized by the various national amateur radio organizations, such as the RSGB in Britain and the ARRL in the USA, but there are many others which are organized by radio magazines and by individual amateur radio clubs. Here we shall look at the requirements for some of these awards.

Usually an operating award will require that the station claiming the award shall have worked a certain number of amateur stations either in different countries, zones or areas of one country. Some local awards may be for working stations in a particular town or perhaps working members of a particular club or society. For many awards, QSL cards must be obtained as proof of contact and these are sent off with the claim for the award. For awards arranged by small clubs, a certified copy of the log entries is often sufficient.

We shall now look at the requirements for some of these awards available from various countries.

French awards

The French society REF has awards for working French provinces and French Departments. More details of these awards are available from REF, 2 Square Trudaine, 75009 Paris, France.

For the DPF (Diploma Provinces France) award, stations in 22 provinces must be worked and confirmed. There are separate awards for phone and CW contacts and a special award if all contacts are on the same band. There is also a five-band award for working 22 different provinces on five bands.

The 22 provinces of France are: Alsace, Aquitaine, Auverne, Basse-Normandie, Bourgogne, Bretagne, Centre, Champagne, Corse, Franche-Comte, Haute-Normandie, Languedoc-Rousillon, Limousin, Lorraine, Midi-Pyrenees, Nord, Pays de Loire, Picardie, Poitou-Charentes, Provence Cote d'Azur, Ile de France, Rhone-Alpes.

For the DDFM (Diploma Departments France Metropolitan), stations from 40 different departments must be worked and confirmed on one band. Extra stickers are available for each additional 10 departments worked and there is a special sticker for working all 95 departments. There is also a five-band award which requires a total of 300 departments on five bands with at least 10 on each band.

There is also a French-speaking universe award, called the DUF, for working stations in French-speaking countries, which are mostly former French colonies. This comes in four levels, with the lowest level requiring five countries on three continents. Higher levels have eight countries on four continents, 10 countries on five continents and 20 countries on six continents. There is also a five-band version of this award.

German awards

The Deutscher Amateur Radio Club (DARC) in Germany offers a number of operating awards.

The main award is the DLD, which involves working a number of German districts which are each defined by a DOK number. The basic awards are the DLD100 and DLD200 for contacts with 100 and 200 DOKs respectively on the 80-m band. Amateurs outside Europe can also use 10, 15 and 20 m for their contacts. There are also DLD100/40 and DLD200/40 awards for working 100 and 200 DOKs on the 40-m band.

There is a DLYL award for working 25 German YL operators. Stations outside Europe need only 10 German YL contacts for this award.

The Worked All Europe award involves making contacts with European countries on several different bands. One point is awarded for each country on one band, and individual countries may be worked on up to five bands. For stations outside Europe, countries worked on 80 and 160 m count 2 points each.

The lowest award is WAE3, which requires 40 countries and 100 points. For the WAE2 award, 50 countries and 150 points are required. The highest award is the WAE1, for which 55 countries and 175 points are needed. There are two versions of the WAE awards. One is for two-way CW contacts only and the other is for all contacts by phone.

The DARC also has awards for FAX, RTTY and SSTV operation. For the basic FAX award, a total of five European countries and 10 different European callsign prefixes must be worked and confirmed using FAX. Higher awards are available for 10 countries (20 prefixes) and 20 countries (40 prefixes). For the SSTV awards, each two-way SSTV contact counts as 1 point. The basic SSTV diploma requires 25 points and stickers are available for 50 and 75 points. For holders of the 75-point award, the SSTV trophy can be gained by reaching 100 points.

American awards

Most of the major awards from the USA are organized by the American Radio Relay League (ARRL). Many other awards are organized by individual clubs and magazines.

Perhaps the most well-known award is the Dx Century Club (DXCC). This award is for confirmed contacts with 100 different countries as defined in the current DXCC countries list. There are 12 different versions of the DXCC, depending on the mode of operation or the band used.

The simplest DXCC is the mixed-mode one, where contacts may be made using any mode on any band except 10 MHz. More difficult are the single-mode versions, which are for all-CW, all-phone or all-RTTY contacts. Single-band DXCCs can be gained for making all QSOs on one band, which may be 160, 80, 40, 10, 6 or 2 m. There is also a five-band DXCC award which requires 100 countries worked on each of five bands (500 QSOs in all). For the five-band award the new WARC bands (10, 18 and 24 MHz) are not valid. There is also a satellite DXCC award for working 100 countries via amateur radio satellites.

The Worked All States (WAS) award is issued by the ARRL for making confirmed contacts with stations in all 50 states of the USA (including Hawaii and Alaska). Contacts may be made using any mode on any band except 10 MHz. There are also special awards available for all RTTY and all SSTV contacts. Another special award is for making all contacts via OSCAR satellites. Stickers are available for single-mode operation such as all-CW, all-SSB and for all contacts via packet radio. There is also a five-band WAS award where all 50 states must be worked on each of five bands.

The Worked All Continents (WAC) award is issued by the IARU for stations which have made contact with each of the six continental areas of the world (Europe, Asia, Africa, North America, South America and Oceania).

The basic WAC may be obtained by making the contacts on any band using any mode. Single-mode certificates can be gained by making all of the contacts using one mode, which may be CW, phone, RTTY, FAX or SSTV. Another special WAC is available for making all contacts via satellite. There is also a five-band award for working all continents on each of five bands.

Endorsement stickers can be gained for QRP (5 W maximum output), 1.8-MHz band, 3.5-MHz band and 50-MHz band operation. Note that contacts on the 10-, 18- and 24-MHz bands are not valid for WAC.

The Worked All Zones (WAZ) award is organized by *CQ Magazine* and involves working stations in all 40 of the CQ Dx zones of the world as defined by *CQ Magazine*.

Awards may be endorsed for single-mode operation using CW, SSB or RTTY. There are also single-band endorsements for making all contacts on one HF band, which may be 160, 80, 40, 20, 15 or 10 m. There is also a five-band WAZ award for working all 40 zones on each of five different bands (200 contacts).

CQ Magazine also offers a (Worked Prefixes) WPX award for making confirmed contacts with a certain number of amateur call prefixes. The call prefix is the first letter and number of the call. Thus G3, G6, GM4, GW0 and so on all count as separate prefixes. The basic award is for working 400 different prefixes using mixed modes on any bands. Single-mode awards are available for working 300 prefixes using CW or 300 using SSB. Endorsements can be added for each extra 50 prefixes worked.

Another CQ award is the CQ Dx award, which is similar to the DXCC since it requires QSOs with 100 different countries using two-way CW or SSB. Extra endorsements are available for working 150, 200, 250, 275, 300 and 320 countries. There are also special endorsements to the basic award for QRP, mobile, SSTV and satellite contacts, where 50 countries using the special mode are required.

The USA Counties award is also run by *CQ Magazine*, and involves working stations in 500 different counties within the USA. Higher versions of the award are for 1000 counties in 25 states and 1500 counties in 45 states. Further awards are available for 2000, 2500 and 3000 counties with all 50 states worked as well.

There is an ITU Zones award which requires contacts with 40 different ITU zones. Higher grade awards are also available for working 50, 65, 75 and 90 zones.

UK awards

The Commonwealth Century Club award requires confirmed contacts with 100 stations in different British Commonwealth call areas since 1984. Successful applicants receive a plaque commemorating the award. Contacts on 10-, 18- and 24-MHz bands do not count for this award.

There is also a five-band version of this award. This comes in five classes, with the highest being the supreme class award, which requires contacts with 500 Commonwealth stations on five bands. Call areas can be counted only once per band. There are also class 1 and 4 awards, which require 450, 400, 300 and 200 contacts respectively. Class 4 requires a minimum of 30 different call areas on each band.

The Commonwealth call areas are:

A2 Botswana, A3 Tonga, C2 Nauru, C5 Gambia, C6 Bahamas, G England, GD Isle of Man, GI Northern Ireland, GJ Jersey, GM Scotland, GU Guernsey, GW Wales, H4 Solomon Islands, J3 Grenada, J6 St Lucia, J7 Dominica, J8 St Vincent, P2 Papua New Guinea, S2 Bangladesh, S7 Seychelles Islands, T2 Tuvalu, T3 Kiribati, V2 Antigua and Barbuda, V3 Belize, V4 St Christopher and Nevis, V8 Brunei, VE1 Maritimes, VE2 Quebec, VE3 Ontario, VE4 Manitoba, VE5 Saskatchewan, VE6 Alberta, VE7 British Columbia, VE8 North West Territories, VK0 Heard Island, VK1 Canberra, VK2 New South Wales, VK3 Victoria, VK4 Queensland, VK5 South Australia, VK6 Western Australia, VK7 Tasmania, VK8 Northern Territories, VK9N Norfolk Island, VK9X Christmas Island, VK9Y Cocos Island, VK9Z Willis Island, VO1 Newfoundland, VO2 Labrador, VP2 Windward and Leeward Islands, VP5 Turks and Caicos Islands, VP8 Falkland Islands, VP9 Bermuda, VQ9 Chagos Island, VR6 Pitcairn Island, VS5 Brunei, VS6 Hong Kong, VU India, VY1 Yukon, YJ Vanuatu, Z2 Zimbabwe, ZB Gibraltar, ZC4 Cyprus, ZD7 St Helena, ZD8 Ascension, ZD9 Gough Island, ZF Cayman Islands, ZK1 North Cook Island, ZK1 South Cook Island, ZK2 Niue Island, ZK3 Tokelau Island, ZL1 Auckland, ZL2 Wellington, ZL3 Christchurch, ZL4 Dunedin, ZL5 Antarctica (NZ), ZL7 Chatham Island, ZL8 Kermadec Island, ZL9 Auckland and Campbell Islands, 3B6 Agalega Island, 3B7 Brandon Island, 3B8 Mauritius, 3B9 Rodriguez Island, 3D2 Fiji, 3D6 Swaziland, 4S7 Sri Lanka, 5B4 Cyprus, 5H Tanzania, 5N Nigeria, 5W1 Western Samoa, 5X Uganda, 5Z4 Kenya, 6Y5

Jamaica, 7P8 Lesotho, 7Q7 Malawi, 8P6 Barbados, 8Q Maldive Islands, 8R Guyana, 9G1 Ghana, 9H Malta, 9J Zambia, 9L Sierra Leone, 9M2 West Malaysia, 9M8 Sarawak, 9V Singapore, 9Y4 Trinidad and Tobago.

The Worked ITU Zones award requires contacts with stations in at least 70 of the 75 ITU zones. Stations qualifying for this award receive an engraved plaque. There is also a five-band version of this award.

The Islands on the Air (IOTA) award is made to stations working amateur stations located on islands around the world. There are a whole series of different awards. Some are for islands in various regions of the world. These include IOTA Arctic Islands, IOTA Antarctica, IOTA Europe, IOTA Africa, IOTA Asia, IOTA North America, IOTA South America and IOTA Oceania. There are also IOTA Century Awards for working 100, 200, 300 and 400 islands.

Full details of the qualifying islands and the numbers required for the various awards can be obtained from the RSGB, who administer this award.

There are many other awards organized by national radio societies and by individual clubs. Often the details of these are given in amateur radio magazines.

23

Power supplies

An important part of all electronic equipment is the power supply, which in most cases is derived from a 110- or 240-V AC mains supply. In many mains supply systems a transformer and rectifier system operating at the mains frequency of 50 or 60 Hz is used to generate the various voltage supplies needed by the equipment. In recent years an alternative system, known as a switch mode power supply, has become popular. In this type of supply a high-frequency power oscillator is driven from the mains input so that the transformer and rectifier system works at a frequency of perhaps 100 kHz. This type of supply is generally more compact and cheaper to produce than a mains frequency supply.

For portable and mobile operation where no mains supply is available, the supply voltages for the equipment are derived from batteries. A typical battery consists of a number of individual electrical cells connected together. Various types of construction are used for the cells in the battery. All cells rely on chemical reactions to generate electricity. Some types, known as primary cells or dry cells, produce electricity for a period until the chemical reaction ceases, and then they are discarded. Other types of cell, known as secondary cells, can be recharged by passing an electric current through them which causes the chemical reaction to be reversed.

Primary cells
There are several varieties of primary cell which may be used in radio equipment. For powering the equipment, the two main types are the zinc carbon type and the manganese oxide type. Other types of primary cell which may be used for long-term light-duty operation for running clocks or providing memory back-up are the mercury cell, silver oxide cell and the lithium cell.

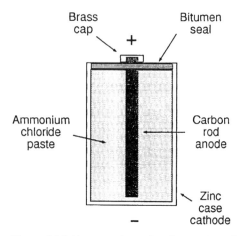

Figure 23.1 Construction of a zinc carbon dry cell

Zinc carbon cells

This type of cell is based on the early Leclanche cell and its basic construction is shown in Figure 23.1. A zinc outer casing acts as the negative electrode of the cell and the carbon rod in the centre is the positive electrode. The space between the electrodes is filled with ammonium chloride paste, which acts as the electrolyte. A new cell produces a voltage of about 1.54 V between the electrodes. When current is drawn, a chemical reaction takes place which oxidizes the zinc electrode. After a period of discharge, the electrolyte becomes less effective and the internal resistance of the cell rises so that, when delivering current, the voltage falls to about 0.8 V and the cell is considered to have reached the end of its life. At this point the cell is usually discarded.

A number of different sizes of zinc carbon cell are in common use. These have a variety of type numbers but are often described by the American size letter codes. The largest cell is the D type, which is commonly used for torches and other heavy-current applications. The next largest is the C cell, which is used for medium-current applications such as cassette recorders and radios. A smaller cell is the AA size, which is widely used for lighter current applications, such as portable radios and TV remote controls. There is also a smaller AAA size cell, which is sometimes used in miniature equipment and remote control handsets.

It is also possible to obtain batteries which contain several cells in a single package. One common type is the PP3 battery, which is built in a series of layers and gives an output of 9 V, which is ideal for use in small portable radios. There is also a larger version of this layer-type battery known as the PP1.

The capacity of a battery to deliver current is usually specified in ampere hours (AH), which defines the life of the cell in hours when delivering a specified current. For zinc carbon cells this is usually based on intermittent use for 4 hours per day.

Typical capacities for standard zinc carbon cells and the current level during the 4-hour discharge are:

Cell size (mA)	Capacity (AH)	Current (mA)
AAA	0.520	5
AA	1.020	20
C	1.920	20
D	5.240	40
PP3	0.2710	2

The life of the cell will be shortened if heavier or more continuous current demands are made upon it.

Some chemical reaction takes place within the cell even when it is not delivering current, so that if left unused the cell will eventually become discharged. This type of cell therefore has a definite, although long, shelf life. As the chemical reaction progresses, zinc is removed from the zinc electrode and eventually pinholes may develop which allow the electrolyte to leak out. This can cause damage inside a piece of electronic equipment. On equipment which is used infrequently, it is wise to check the state of the cells from time to time and to replace any that show signs of leakage, which shows as a white deposit on the outside of the zinc case or a soapy liquid on the outside of the cell.

A variation on the basic zinc carbon cell is the high-power version, which uses an improved internal construction to provide a higher current whilst maintaining the output voltage of about 1.5 V. These cells are often enclosed in a steel case and sealed to prevent leakage of electrolyte.

Another variation is the super power cell, which provides even higher current output and is intended for driving equipment containing motors such as a cassette recorder or a motor-driven toy. These cells often use a manganese dioxide positive electrode and zinc chloride as the electrolyte. The life of this type of cell is also longer than that of the standard-type cell.

It is in fact possible to reactivate a zinc carbon cell by using a recharging process in which current is passed into the cell from a cheaper unit. This is most effectively done by using a circuit similar to that shown in Figure 23.2. Here the cell is charged during the positive half-cycle pulse and then a discharge at about a tenth of the charging current occurs during the negative half-cycle. The charging current must be kept low, since the reversal of the chemical reaction within the cell usually produces some gas and the cell will heat up. If the current is too high, the gas pressure inside the cell can build up rapidly and will destroy the cell by blowing out the seal at the top of the

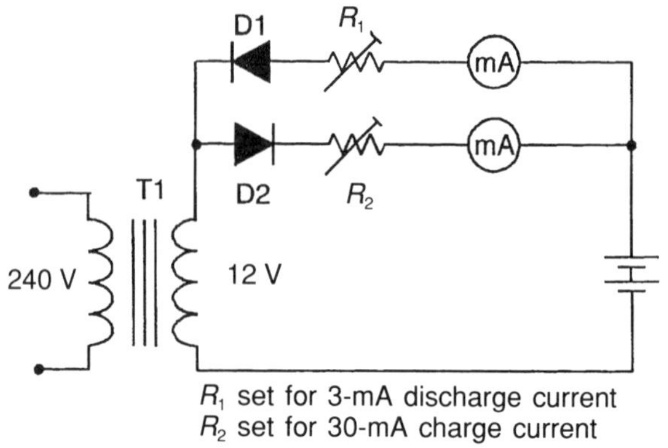

R₁ set for 3-mA discharge current
R₂ set for 30-mA charge current

Figure 23.2 Basic circuit for recharging dry cells

cell, so appropriate precautions must be taken. With care, cells can be reactivated perhaps 10 or 30 times. Unfortunately the zinc does not replate back into the same place on the cell case, and eventually holes will develop, causing the cell to start to leak electrolyte.

The high-power and super power sealed cells are not suitable for recharging in this way since the internal gases cannot escape and a violent explosion could occur if the internal pressure became high enough to break the steel case.

Alkaline cells

An alternative type of primary cell which gives high-current capability and long life uses a manganese dioxide positive electrode and a powdered zinc negative electrode with potassium hydroxide as the electrolyte.

This type of cell gives an output voltage of 1.5 V and has about twice the capacity of a similar-size zinc carbon cell. The main advantage of this type of cell is that it can deliver quite heavy current without serious loss of output voltage, which makes it ideal for applications where there is an intermittent heavy current demand. These cells are available in same-size packages as the zinc carbon types.

Typical capacity ratings for alkaline batteries are:

Cell size (mA)	Capacity (AH)	Current (mA)
AAA	0.965	5
AA	2.020	20
C	6.220	20
D	12.440	40
PP3	0.5310	2.5

The Duracell range of alkaline batteries give about 25% greater capacity than other types.

No attempt should be made to recharge this type of battery.

Lithium cells

One type of primary battery that has found its way in amateur radio equipment is the lithium thionyl chloride battery. Unlike other primary cells, the lithium cell generates a potential of 3 V. These batteries are expensive but offer a very long life with a stable output voltage. Lithium cells are widely used to provide a standby supply for electronic memory chips so that stored information can be maintained whilst the supply power for the equipment is turned off. They are widely used in amateur radio transceivers and scanner receivers to maintain memory contents.

No attempt should be made to recharge a lithium battery, since this will cause hydrogen to be vented off and a violent explosion could result. Care should also be taken not to short-circuit this type of battery.

Mercury and silver oxide cells

Two further types of primary cell which may be encountered are the mercuric oxide and silver oxide cells. The mercuric oxide cell generates 1.35 V and the silver oxide type 1.5 V. These cells are used for applications where very long life and small size are required. Most types are in button-type cases and they are widely used in watches, clocks, calculators and some test equipment.

Secondary cells

In a secondary or rechargeable cell the output voltage is again produced by a chemical reaction. When the battery has been discharged, the chemical reaction can be reversed by passing a current through the cell from an external charger unit. This charge and discharge cycle may be repeated many hundreds of times before the battery reaches the end of its useful life. Eventually the capacity of the cell to hold the charge diminishes and the need for recharging becomes more and more frequent.

Lead acid battery

This type of battery consists of two sets of lead plates immersed in a sulphuric acid solution. A fully charged cell has a typical terminal voltage of about 2.2 V, and this falls to around 1.8 V when the cell is fully discharged. The most familiar version of this type of battery is that used in automobiles. This typically consists of six cells connected in series to give a battery voltage of 12 V, and is designed to give very high current for a short period to drive a starter motor.

The lead acid cell can deliver heavy current and gives a high charge capacity in a relatively small space. For portable equipment there is a variant of this type of cell which is fully sealed and uses an electrolyte in the form of a gel.

Lead acid cells are usually rated based on a current that will discharge the cell in 10 hours. For use with radio equipment, a variety of small lead acid batteries is available with voltages of 6 V or 12 V and capacities ranging from 1 AH to about 40 AH. Since these batteries are sealed, they should not be charged using a battery charger intended for car batteries, which are able to vent off gas during the charging process. The sealed batteries are usually trickle charged at relatively low currents.

Nickel cadmium cells

An alternative to the lead acid cell is the nickel cadmium or Nicad cell. This type of cell has a nickel hydroxide positive electrode and a cadmium negative electrode, whilst the electrolyte is potassium hydroxide. The cells normally have a vent in the top seal, which allows any excess gas pressure built up in the cell during charging or excessive discharge to be vented off, after which the vent reseals itself.

The Nicad cell has a high storage capacity and low internal resistance, so that it can deliver high output currents whilst maintaining its output voltage. Unlike the zinc carbon type, a Nicad cell has an output voltage of only 1.2 V. Nicads are available in the same popular sizes as primary cells. For some amateur equipment, such as handheld transceivers, special battery packs may be used rather than standard cells.

The capacity of typical Nicad cells is as follows:

Cell size	Capacity (AH)
AAA	0.18
AA	0.5
AA*	0.6
C	1.5
C*	2.0
D	1.5
D*	4.0
PP3	0.11

The types marked with an asterisk are industrial versions which have a higher capacity than the equivalent-size commercial cells. It will be noted that, in general, the Nicad cells have a lower capacity than the equivalent zinc carbon type, but of course they can be recharged many times.

Nicad cells are recharged at a constant current for a period of 14–16 hours at a current of about one-tenth the ampere hour capacity. They can be charged at higher current for a shorter period of time if desired, so that an AA cell could be charged at 500 mA for a period of 1.25 hours provided it were fully discharged at the start of the charging process. A typical Nicad cell may be recharged 2000–3000 times before it needs to be replaced. Most manufacturers guarantee that the cell can be cycled at least 500 times.

One problem that can occur with Nicads is that if they are only partly discharged and then topped up again by the charger, they will develop a memory effect in which they no longer produce the full discharge capacity. This can be reversed by fully discharging and recharging the cell a few times. For best results the cell should be fully discharged before it is recharged, since this will avoid the memory effect.

Linear mains supplies

The basic blocks of a mains power supply are as follows. A transformer converts the incoming mains voltage to a number of separate AC signals of the desired voltages. This is followed by one or more rectifiers, which convert the AC voltages into DC. Following each rectifier, there is usually a filter to remove AC ripple from the DC output, and there may also be a voltage regulator to produce a stable DC voltage under varying load conditions.

Rectifiers

To convert the AC signal in a power supply into a DC voltage, one of a number of rectifier configurations may be used. The simplest scheme is the half-wave rectifier shown in Figure 23.3. Here the rectifier diode allows the positive half-cycle of the input waveform to pass through, but blocks the negative half-cycle, when the anode of the diode is driven negative relative to the cathode. The output consists of a series of half-cycle pulses. The capacitor after the rectifier is referred to as the reservoir capacitor, and during each pulse it will charge to the peak voltage of the input sinewave. During the negative half-cycle the capacitor discharges into the load which the supply is driving. If the capacitor is sufficiently large, the output waveform is as shown in Figure 23.4. The small variation in the output voltage is known as the supply ripple and should be a small percentage of the total output voltage

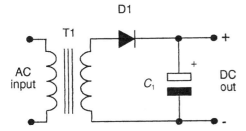

Figure 23.3 Half-wave rectifier power supply circuit

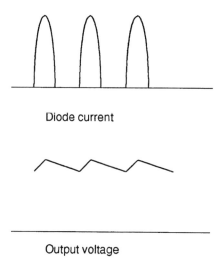

Diode current

Output voltage

Figure 23.4 Output waveform of half-wave rectifier

under full load conditions. In a regulated power supply the regulator circuit will remove any ripple that is present, so that the output becomes pure DC.

In a simple half-wave rectifier system the peak current through the rectifier will be about three times the DC load current, and the capacitor must be capable of handling this level of ripple current. This type of rectifier scheme is generally used for low-current supplies.

A more efficient rectifier system is the full-wave rectifier shown in Figure 23.5. Here a centre tapped winding on the transformer is used and feeds two rectifiers whose outputs are combined at the reservoir capacitor. In this case diode D1 conducts during one half-cycle of the AC input, whilst the second diode conducts during the other half-cycle. This means that for a 50-Hz mains supply there will be 100 half-cycle pulses applied to the capacitor. Since the

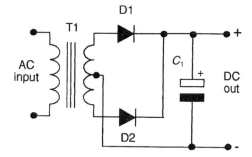

Figure 23.5 Full-wave rectifier power supply circuit

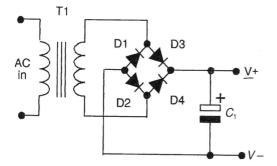

Figure 23.6 Full-wave bridge rectifier circuit

circuit operates on both half-cycles, the peak currents in the transformer are reduced and filtering is easier because the ripple is now at 100 Hz. This means that the reservoir capacitor can be of lower value for the same ripple voltage output and carries a lower ripple current.

One disadvantage of the full-wave rectifier scheme is that a centre tapped winding is needed on the transformer, which tends to make it more expensive. An alternative method of obtaining full-wave operation is to use a rectifier bridge as shown in Figure 23.6. Here diodes D1 and D4 conduct during the positive-going half-cycle, whilst D2 and D3 conduct on the negative half-cycle. The output is the same as for a full-wave rectifier circuit but because there are effectively two diodes in series in the conducting path the voltage drop across the rectifier unit is doubled. Most modern high-current power supplies use a bridge rectifier system.

Another useful rectifier circuit is the voltage doubler shown in Figure 23.7. Here on one half-cycle capacitor C_1 charges via diode D2. On the next half-cycle diode D1 conducts, charging C_2. The output voltage is the sum of these

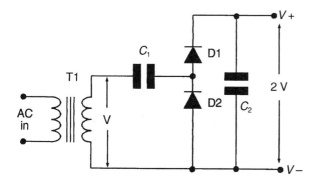

Figure 23.7 Voltage doubler circuit

two voltages, so it becomes double the input AC voltage. This process can be repeated to form a voltage multiplier which can produce a very high voltage from a much lower AC input. This type of supply is often used for producing the high voltage needed for a cathode ray tube, where the current required is small.

Voltage regulation

The simplest scheme for providing a stable output voltage uses a Zener diode. This is a diode which operates in its breakdown region, where if the voltage rises the diode current increases rapidly. A series resistor is used to limit the current in the circuit. When there is no load across the output, enough current will flow through the diode so that the voltage across it is its Zener voltage. When a current is drawn by the load, the diode current falls, keeping the output voltage more or less constant. Zener diodes come with a range of operating voltages from 3.3 V up to perhaps 100 V.

A Zener diode stabilizer is generally used for relatively low-current supplies and is often used to provide stable reference voltages within circuits. Zener diodes can be used to stabilize the supply to individual stages in a circuit where the operating voltage is critical. This can be useful in a circuit designed to operate from a battery supply, since it allows a fairly large variation in battery voltage without altering the stable supplies within the unit.

Series regulators

A common regulator scheme for use in high-current supplies is the series regulator shown in Figure 23.8. Here the series transistor TR1 is effectively acting as an emitter follower, and the output voltage is governed by the base drive voltage relative to the negative supply rail. Transistors TR2 and TR3 act as an inverting amplifier. In this circuit the voltage from RV_1 is compared with the voltage across the Zener diode D1. If the output voltage is high, then transistor TR3 will conduct more than TR2, and the collector voltage of TR3 falls. This in turn reduces the voltage applied to the base of TR1 and hence reduces the output voltage of the supply. Thus any variation in output voltage due to changes in load current are cancelled out to give a stable output voltage. The output voltage level is adjusted by varying RV_1.

For common voltage supplies such as a 5-V logic supply, it is convenient to use one of the fixed-voltage integrated circuit regulators such as the 7805. This device contains all of the circuitry for a series-regulated supply in a single integrated circuit package which has just three terminals for input, output and ground. This type of device usually contains additional circuitry which provides a current limiting action, so that in the event of a short-circuit occurring the circuit limits the current to a safe level. A typical regulator such as the 7805 will produce a 5-V output with a tolerance of perhaps 0.25 V over a current range from 0 to 1 A. The input voltage normally needs

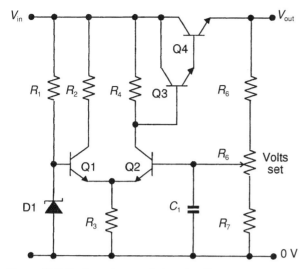

Figure 23.8 Basic series regulator circuit

to be about 1–2 V more than the desired output at maximum load current. Because the regulator contains a high-gain amplifier, it is important to decouple the output line with a 100-nF ceramic capacitor mounted close to the regulator output pins. This prevents high-frequency oscillation within the regulator chip. The commonly available fixed regulators are the 7805 (5 V), 7812 (12 V) and 7815 (15 V) types, which provide a 1-A output capability. There are also lower power versions, such as the 78M series for 500-mA output and the 78L series for 100-mA output. For negative-voltage-regulated supplies, the 7905, 7912 and 7915 may be used and these also have 79M and 79L versions for lower current output operation.

It is also possible to obtain integrated circuit voltage regulators where the output voltage can be adjusted to any desired present level. A typical example is the LM317, which can provide up to 1.5 A at output voltage from +1.2 V to +33 V. The input voltage should be 2–3 V more than the output voltage.

Switch mode supplies

One disadvantage of the conventional mains power supply is that the transformer has to have a large core in order to produce sufficient inductance so that the magnetizing current is kept to a low value. This is the current that flows in the transformer when no load is applied to the power supply output. Another problem is that very large value capacitors are needed for ripple filtering after the rectifier stage.

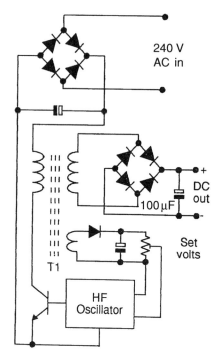

Figure 23.9 Block diagram of switch mode power supply

An alternative approach to the design of power supplies is the switch mode supply. In this type of unit the raw mains input is rectified to produce a high DC voltage which feeds a high-frequency oscillator, running at about 100 kHz, and a power amplifier stage as shown in Figure 23.9. The output of this power amplifier is fed to a transformer. The secondary windings of the transformer feed rectifiers and filters with the same general arrangement as in a conventional low-frequency supply.

Because the power input to the transformer is at high frequency, the primary inductance of the transformer can be very much smaller than in a 50-Hz supply for the same magnetizing current. The transformer core can be much smaller and may be made from a ferrite material rather than conventional iron. The high-frequency signal is also much easier to filter after it has been rectified, since the capacitors needed can be lower in value for the same current level.

Regulation of the output voltage from this type of supply can be achieved by using feedback from the DC output to control the on/off ratio of the pulses generated by the oscillator, which effectively controls the mean power input to the transformer primary. By varying the amount of feedback, the DC output voltage from the supply can be preset to a desired level. In most switch mode power supplies, the oscillator and mark–space control circuits are contained in an integrated circuit. Conventional linear regulators may also be used in the DC output lines of a switch mode power supply to provide overload protection and to set individual output voltage levels. Thus the basic supply might produce a loosely regulated 15-V output, and series regulators are then used to produce, say, 12- and 5-V stabilized outputs.

One point to note in this type of supply unit is that the input rectifier, oscillator and power amplifier circuits are directly linked to the mains supply. The transformer provides isolation between the DC output circuits and the mains input, so that the equipment can be earthed on the secondary side of the transformer for safety. In some switch mode supplies the feedback from the output DC to the oscillator control circuits is made via an opto-coupler, so that there is no DC path between the primary and secondary circuits of the transformer.

Switch mode supplies are widely used in TV receivers and in computers. One point to note is that some switch mode supplies will not operate properly with no load current. To prevent high-frequency radiation from the switch mode supply circuits, the unit will often be mounted inside a screening box and all input and output lines are usually filtered to remove any high-frequency signals.

24

Electronic components

Although many modern electronic systems use complex integrated circuits to replace most of the components of the circuit, it is still necessary to use discrete components such as resistors, capacitors, diodes and transistors when building electronic projects. For the amateur constructor it is helpful to understand the characteristics of the various components used and to recognize their circuit symbols.

Resistors

A resistor, as its name implies, is a component which provides resistance to the flow of electric current. The value of resistance is specified in a unit called the ohm (Ω), where a resistance of 1 Ω will produce a voltage difference of 1 V when 1 A of current is flowing through it. Thus resistance in ohms is given by:

$$R = \frac{V}{I}\ \Omega$$

where V is the voltage drop across the resistor when a current of I amperes is flowing through it.

There are several different forms of construction for resistors, including carbon composition, carbon film, metal oxide and metal wire types. The type used depends upon the application.

Carbon composition resistors consist of a rod made from a mixture of carbon and insulating material. The mixture used and the dimensions of the rod determine the value of resistance. The value of these resistors is not particularly accurate and is usually specified to be within perhaps 20% of the nominal value. This type of resistor is not widely used today but may be found in old equipment.

Carbon film resistors are perhaps the most common type in use. To produce the resistance element, a thin film of carbon is deposited on a ceramic rod and the whole resistor is then coated with a layer of insulating material. These resistors are usually produced with a value of tolerance of 5%. The carbon film resistor provides good stability with temperature changes. Typical values of temperature for these resistors ranges from 150 to 850 parts per million per degree Celsius. These resistors also have good low-noise characteristics, which is important for use in high-gain amplifiers and RF applications.

Carbon film resistors come in a range of physical sizes, giving power dissipation ratings of 0.125, 0.25, 0.5, 1.0 and 2.0 W. The widest range of values is usually available in the lower wattage rating resistors.

A variation of the film resistor uses a metal film instead of carbon. This gives much better precision for the resistance value and also very high temperature stability. The temperature coefficient of this type of resistor is 50–100 parts per million per degree Celsius. The metal film resistors are usually available in 0.125-, 0.25- and 0.5-W ratings. The value tolerance for this type of resistor is within 1% of the nominal value. This type of resistor should be used wherever the value of resistance required is critical to the operation of the circuit.

In any circuit the resistor will absorb some power from the current flowing through it, and this causes the temperature of the resistor to rise. Carbon and metal film resistors are typically used for applications where the power dissipated by the resistor is less than 0.5 W. Some carbon resistors can be obtained with power ratings up to 2 W, but their size increases in proportion. For high-power applications a wirewound resistor should be used. The wirewound resistor consists of a length of resistance wire wound onto a ceramic former and coated with a vitreous enamel or silicone cement insulation. The resistance wire is usually a nickel alloy and is of small diameter, so that it produces a high value of resistance for a given length. This type of resistor runs fairly hot when in operation, so it should be placed in a position where there is a good circulation of air to carry away the heat.

Wirewound resistors intended for mounting on a circuit board in the same way as a carbon or metal film type come in power ratings of 2.5 and 6 W. For higher power applications the resistor is mounted in a ceramic case, and power ratings of 4, 7, 11 and 17 W are available. These resistors are designed to stand vertically on a circuit board to give good air circulation. Higher power wirewound resistors are built into an aluminium case, which is designed to be bolted to a heatsink. These resistors come with power ratings of 10, 25, 50, 100 and 200 W. When run without a heatsink, these resistors will handle about half their maximum power rating.

Wirewound resistors have very good temperature stability and their value tolerance is usually 5%. It is also possible to obtain low-wattage precision wirewound resistors for use in instrumentation. These resistors are usually rated at 0.3 W and come with a value tolerance of 0.1%.

For some applications in logic systems a number of resistors of the same value may be required to act as pull-up or pull-down resistors or as current limiters in led circuits. A range of resistor networks is available in integrated circuit style single or dual in line (DIL) packages. This has the advantage of reducing the amount of space needed on the printed circuit board.

Resistor arrays may have up to eight separate resistors, where both ends of each resistor are brought out to separate pins. Other types have one end of all of the resistors joined to a single pin, and here up to 15 resistors may be included in the package. Packages may be 8–10-pin SIL types or 14-pin and 16-pin DIL types. The resistors in the package usually have a value tolerance of 2% and a power rating of 0.125 W. Values range from 33 Ω to 100 Ω, with the value sequence 1.0, 2.2, 3.3 and 4.7 in each decade.

Potentiometers

For many applications a variable resistor or a potentiometer is required. These consist of a resistor track along which a contact can be moved. For volume controls and other manually operated controls the track is circular and the contact is mounted on a shaft which extends through the equipment front panel. For other applications where the control only needs occasional adjustment a preset potentiometer is used. Some presets use circular track and are designed to be adjusted by using a screwdriver. Some types use a straight track with a sliding contact which can be moved along it.

For most applications a carbon track potentiometer can be used. These are available with linear or logarithmic tracks. The logarithmic types are used for audio volume controls, where a log law is desirable. For other applications a linear potentiometer would be used. Potentiometers with plastic film or cermet tracks are available for applications where long life and accuracy are required.

Preset carbon track potentiometers are available in two types for either horizontal or vertical mounting on a printed circuit board. On these presets a single turn of the adjustment screw moves the wiper from one end of the track to the other. For more precise applications preset potentiometers with cermet tracks may be used. These are also physically smaller than the carbon track types and are usually designed with 0.1-inch lead pitch. For more precise setting of the preset, a multi-turn-type cermet potentiometer may be used. These require 10–25 turns of the adjusting screw to move the slider from end to end of the track. Most multi-turn presets have a slipping clutch to prevent damage when an attempt is made to drive the control beyond the end of its track.

Resistance calculations

When two resistors R_1 and R_2 are connected in series, the combined value is given by:

$$R = R_1 + R_2 \ \Omega$$

where R is the combined resistance and R_1 and R_2 are the individual resistor values in ohms. This can be extended to any number of resistors, where the combined value is the sum of the values of all of the resistors in the series chain.

When two resistors R_1 and R_2 are connected in parallel, the effective resistance is given by:

$$R = \frac{R_1 R_2}{R_1 + R_2} \; \Omega$$

where R_1 and R_2 are the resistor values in ohms. If the resistors are equal, the value is half that of the individual resistors.

The power dissipated in a resistor is given by:

$$P = I^2.R \text{ Watts}$$

where I is the current flowing through the resistor in amperes and R is the resistance in ohms. Alternatively, the power can be calculated from the voltage across the resistor by using:

$$P = \frac{V^2}{R} \text{ Watts}$$

where V is the voltage across the resistor in volts and R is resistance in ohms.

Preferred values

For resistors a series of preferred values is normally used. For the wider tolerance components a subset of the full series is generally used. The three common sets of values in use are known as the E6, E12 and E24 series.

E6 series: 1.0, 1.5, 2.2, 3.3, 4.7 and 6.8.

E12 series: 1.0, 1.2, 1.5, 1.8, 2.2, 2.7, 3.3, 3.9, 4.7, 5.6, 6.8 and 8.2.

E24 series: 1.0, 1.1, 1.2, 1.3, 1.5, 1.6, 1.8, 2.0, 2.2, 2.4, 2.7, 3.0, 3.3, 3.6, 3.9, 4.3, 4.7, 5.1, 5.6, 6.2, 6.8, 7.5, 8.2 and 9.1.

Thus in the E6 series the resistor values would be 10, 15, 22, 33, 47, 68, 100, 150, 220 and so on.

Resistor codes

In circuit diagrams the resistor value is given as a numerical code rather than as a full value. The system uses a letter combined with the preferred value. The letter replaces the decimal point and indicates the value multiplier as follows:

R = ohms
k = kilohms
M = megohms

Thus 4R7 represents a value of 4.7 Ω

47R = 47 Ω
k47 = 470 Ω
4k7 = 4700 Ω
47k = 47 kΩ
4M7 = 4.7 MΩ

A letter following the value code indicates the resistor tolerance as follows:

F	±	1%
G	±	2%
J	±	5%
K	±	10%
M	±	20%

Many modern resistors are marked with their value using these codes.

Colour codes

Many resistors have the value indicated by means of a colour code rather than by using figures. On modern resistors the colour coding consists of a series of coloured bands as shown in Figure 24.1. The resistance value is given by the first two coloured rings, with the third ring indicating the multiplier factor as follows:

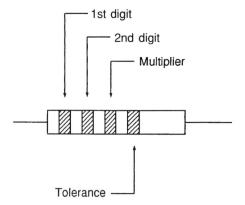

Figure 24.1 Standard colour code scheme for resistors

Colour	Number	Multiplier
Black	0	×1
Brown	1	×10
Red	2	×100
Orange	3	×1000
Yellow	4	×10 000
Green	5	×100 000
Blue	6	×1 000 000
Purple	7	not used
Grey	8	not used
White	9	not used
Silver	–	×0.01
Gold	–	×0.1

A fourth band may be used to indicate tolerance as follows:

Brown	±	1%
Red	±	2%
Gold	±	5%
Silver	±	10%
None	±	20%

Thus a resistor with yellow, purple, orange and silver rings would be a 47-kΩ resistor with a tolerance of 10%.

Recently an alternative coding scheme has appeared, particularly on close tolerance resistors. This uses five bands as shown in Figure 24.2. In this case three bands are used to indicate the value, while the fourth is a multiplier. The fifth band denotes the value tolerance. Thus a resistor with brown, black, black, red for the value bands is a 10-kΩ resistor and not a 1-kΩ type as it might first appear to be.

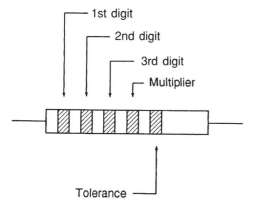

Figure 24.2 Alternative colour code for 1% resistors

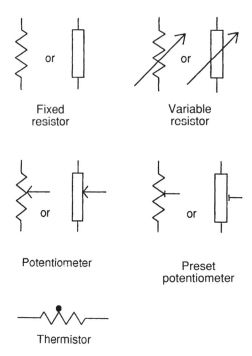

Fixed
resistor

Variable
resistor

Potentiometer

Preset
potentiometer

Thermistor

Figure 24.3 Resistor and potentiometer circuit symbols

Resistor circuit symbols

In most modern circuit diagrams a resistor is shown as a simple rectangular box with the resistance value written alongside it as shown in Figure 24.3, which also shows the symbols for variable and preset resistors and potentiometers. Older circuit diagrams may show resistors as a zigzag line rather than a box.

Capacitors

A capacitor consists basically of two metal plates separated by a thin layer of insulating material. In its simplest form the insulating material may be air. Air dielectric capacitors are widely used as tuning capacitors in receivers and transmitters. These are normally made variable with a set of moving plates which interleave with a set of fixed plates.

For RF applications a mica dielectric capacitor is useful, since it has close value tolerance, high stability and low losses. Usually these are built in layered form with a stack of plates separated by mica sheets and encapsulated in plastic. The value tolerance on typical mica capacitors is 1% or 2%.

For low-frequency work, paper dielectric capacitors were widely used in the past, but today these have been replaced by capacitors using various plastic film dielectrics.

For values of capacitor up to about 10 nF, the polystyrene capacitor provides close tolerance values and high-voltage operation and is physically smaller than a mica capacitor. This type of capacitor is often used for applications where a precise value is needed such as in tuned circuits, filters and timing circuits. The value tolerance is typically 2% with a DC voltage rating of 160 V. The temperature coefficient of this type of capacitor is the order 150 parts per million/°C.

Polyester capacitors are generally useful for AF and low-frequency applications where values from 10 nF to 2.2 μF are required. They are compact and have DC voltage ratings of 100, 250 or 400 V. Value tolerance is usually 20% but the temperature coefficient is 200 ppm/°C.

Polycarbonate capacitors come in a range of values similar to that for polyester types and with voltage ratings of either 160 or 630 V. These capacitors have a better temperature stability than polyester types at around 50 ppm/°C.

Polypropylene capacitors give tolerances of around 5% and voltage ratings of 53 or 160 V with values from 100 pF to 10 hF. They can be used in similar applications to the polystyrene types. High-voltage versions are also available with values from 1 to 470 nF and voltage ratings of 1000 or 1500 V.

Ceramic capacitors consist of a thin plate of ceramic material plated on each side with a metal film. The ceramic material has a very high value of permittivity (K), so that high values of capacitance can be achieved in a small physical space. This type of capacitor is generally a low-voltage device with typical ratings up to perhaps 100 V. One special variant of the ceramic capacitor is the high-voltage type with ratings up to perhaps 10 kV, which is used for decoupling in high-voltage TV picture tube circuits.

Ceramic disc capacitors are generally used for RF decoupling where the actual value of the capacitor is not particularly important and where high capacitance and low lead inductance are desirable.

Electrolytic capacitors use a thin metal oxide film on one of the capacitor plates as a dielectric, and the space between the plates is filled with a layer of insulator containing an electrolyte which acts as a conductor. These capacitors are normally polarized and will only work properly when the polarity of the voltage across the capacitor is correct.

Because the oxide layer is very thin, high values of capacitance can be obtained in a very small package. Usually the electrodes are made in the form of a strip and then rolled up and installed in a circular aluminium can which acts as the negative terminal of the capacitor.

One problem with electrolytic capacitors is that they have a relatively low resistance between the plates, so that at normal working voltage a significant value of leakage current will flow through the capacitor. The main application for electrolytic capacitors is for decoupling at audio frequencies and as reservoir and filter capacitors in power supplies.

Capacitor value codes

Most capacitors have the value marked either directly in pF or µF or by using a coding system similar to that used for resistors. The values usually follow the E6 sequence, although closer tolerance capacitors may have values in the E12 series. The multiplier letters used for capacitor values are:

Picofarad	p
Nanofarad (1000 pF)	n
Microfarad (1000 nF)	µ

Examples:

Marketing	Value
p47	0.47 pF
4p7	4.7 pF
47p	47 pF
4n7	4700 pF
47n	0.047 µF
470n	0.47 µF
4u7	4.7 µF

Values less than 1000 pF are marked in pF, and above 1 µF in µF, with other values marked in nF.

Some capacitors are marked with a two-figure value in pF followed by a third figure which is a power of 10 multiplier.

$152 = 15 \times 10^2 = 1500$ pF or 15 nF
$473 = 47 \times 10^3 = 47\,000$ pF or 47 nF
$104 = 10 \times 10^4 = 100\,000$ pF or 0.1 µF

Capacitance calculations

Two capacitors C_1 and C_2 connected in series produce an equivalent capacitance of:

$$C = \frac{C_1 C_2}{C_1 + C_2} \, \mu F$$

where C, C_1 and C_2 are capacitor values in µF.

Two capacitors C_1 and C_2 connected in parallel produce an equivalent capacitance of:

$$C = C_1 + C_2 \, \mu F$$

where C, C_1 and C_2 are capacitance values in µF.

The reactance of a capacitor is inversely proportional to both the capacitance and the frequency of operation, as follows:

$$X_c = \frac{10^6}{2\,\pi f C}\ \Omega$$

where $\pi = 3.142$, f is frequency in Hz and C is capacitance in μF.

Capacitance of parallel plate capacitor:

$$C = \frac{0.885KA}{d}\ \text{pF}$$

where A is plate area in cm², d is spacing in cm, and K is dielectric constant of insulator (for air insulation, $K = 1$).

Capacitor symbols

The symbols used to indicate capacitors on circuit diagrams are shown in Figure 24.4. Note that for electrolytic capacitors the polarity may be indicated either by a + sign or by the positive electrode being shown as an unfilled bar.

Inductance calculations

Two inductors connected in series give an equivalent inductance of:

$$L = L_1 + L_2\ \mu\text{H}$$

where L, L_1 and L_2 are inductances in μH.

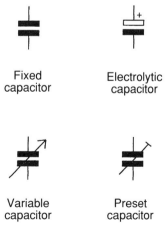

Fixed capacitor   Electrolytic capacitor

Variable capacitor   Preset capacitor

Figure 24.4 Circuit symbols for capacitors

Two inductors connected in parallel produce an equivalent inductance of:

$$L = \frac{L_1 L_2}{L_1 + L_2} \ \mu H$$

where L, L_1 and L_2 are inductance in μH.

The reactance of an inductor is proportional to the inductance and the frequency of operation and is given by:

$$X_1 = 2\pi f L \ \Omega$$

where $\pi = 3.142$, f is frequency in Hz and L is inductance in Henrys.

Inductance of air cored coils:

$$L = \frac{a^2 n^2}{9a + 10b}$$

where

n = number of turns
a = radius of coil (inches)
b = length of coil (inches)
L = inductance of coil (μH)

Number of turns required for inductance $L \ \mu H$:

$$n = \frac{9a + 10b}{La^2}$$

Inductor and transformer symbols

On most circuit diagrams, inductors and the windings of transformers are shown as a coil as shown in Figure 24.5, but on some modern diagrams the windings may be shown as a solid filled bar.

Semiconductor diodes

Most modern semiconductor diodes are of the junction type, in which a p-n junction is created in a silicon bar as shown in Figure 24.6. The junction acts as a rectifier, with the p side acting as the anode and the n side as the cathode. When the anode is positive relative to the cathode the diode conducts, but if the anode is negative relative to the cathode the diode blocks the flow of current.

For general circuit applications such as signal detection, logic and switching functions a small signal diode such as the 1N914 or 1N4148 can be used. These come in a small glass or plastic wire-ended package similar in size to a 0.125-W resistor. For some applications a germanium point contact diode

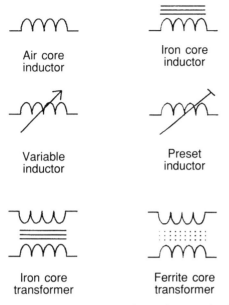

Figure 24.5 Inductor and transformer circuit symbols

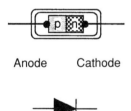

Anode   Cathode

Figure 24.6 Structure of p-n junction diode

may be useful, since it has a lower anode-to-cathode voltage drop when conducting. Signal diodes are generally designed to handle currents from 40 to 75 mA and have reverse voltage ratings of 50–100 V. For very fast switching circuits and high-frequency detectors or mixers, a Schottky barrier diode such as the BAR28 may be used.

For power rectification, diodes with a larger junction area and higher current capacity are used. For applications requiring currents up to 1 A the 1N4000 series diodes may be used, whilst for currents up to 3 A the 1N5400 series are suitable. These two series of diodes come in various voltage ratings from 50 V to 1000 V as follows:

Voltage rating (V)	Diodes	
50	1N4001	1N5400
100	1N4002	1N5401
200	1N4003	1N5402
400	1N4004	1N5404
600	1N4005	1N5406
800	1N4006	1N5407
1000	1N4007	1N5408

These diodes are mounted in similar packages to signal diodes but are larger in diameter. For higher current requirements diodes are usually mounted in metal cases which are designed to be bolted to a metal heatsink to carry away the heat generated within the diode.

For bridge rectifier circuits the four diodes can be obtained in a single plastic package and internally connected as a bridge. Current ratings from 1.5 A to 25 A are readily available. These units are designed to be fixed to a heatsink.

When a junction diode has a negative voltage applied to its anode it will normally block the flow of current, but at some voltage level the junction breaks down and very high currents can flow. This effect is used in the Zener diode to provide voltage stabilization. If a resistor is connected in series with the diode to limit the current to a safe level, the reverse voltage across the diode when in the breakdown region remains virtually constant over a wide range of diode current.

Zener diodes are specified by their reverse breakdown voltage, which may be from 3.3 V up to 75 V with values following the E12 series.

When a junction diode is reverse biased at voltage levels below the breakdown point, the capacitance between the anode and cathode varies with the voltage. Special diodes have been developed to take advantage of this effect and are widely used to replace tuning capacitors in receiver and oscillator circuits. This type of diode is known as a varicap diode. A typical diode for use in an AM radio receiver might have a capacitance range from 15 to 500 pF for a voltage change from 0.5 to 8 V, with values following the E12 series as used for resistors.

When operating, the Zener diode will be carrying a significant amount of current and should be able to absorb the load current from the circuit which it is trying to stabilize. As a result the Zener diode has to dissipate some power. For low-power applications the BZY88C or BZX55C series Zener diodes provide a power dissipation of up to 500 mW, and the voltage rating has a 5% tolerance. For higher power applications the BZX61C or BZX85C series Zeners may be used. These have a power rating of 1.3 W and voltage tolerance of 5%.

If a junction diode is made using gallium arsenide phosphide (GaAsP) and is biased to conduct current, light can be emitted from the junction

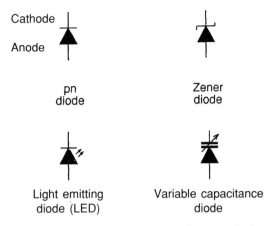

Cathode

Anode

pn
diode

Zener
diode

Light emitting
diode (LED)

Variable capacitance
diode

Figure 24.7 Circuit symbols for solid-state diodes

area. Specially designed diodes are built to maximize this effect and are called light-emitting diodes or LEDs. LEDs can be obtained which emit red, green or yellow light. Some types have two diodes mounted in the same package and can be made to give red, green or yellow light according to the voltage applied to the device. LEDs are widely used for numerical displays on electronic equipment, and individual diodes often act as status indicators. Diodes can also be built which are sensitive to light and produce a voltage when light falls on the diode junction. These are known as photodiodes.

The circuit symbols for the various types of diode are shown in Figure 24.7.

Bipolar transistors

The most common type of transistor is the bipolar npn or pnp junction transistor. These have three electrodes known as the emitter, base and collector. By varying the current flowing between the base and emitter the current between collector and emitter can be controlled so that the transistor can act as an amplifier.

Transistors of the npn type are designed to run with the collector positive relative to the emitter, whilst pnp types have the collector negative to the emitter. The circuit symbols used for pipolar transistors are shown in Figure 24.8.

For general-purpose AF and other low-frequency circuits the BC107, BC108 and BC109 npn types and the BC177, BC178, BC179 pnp types may be used. For RF and fast switching applications, transistors with a much higher cut-off frequency f_T must be used. Transistors for signal audio and switching applications are usually low-power types with a dissipation rating of 300–500 mW. For

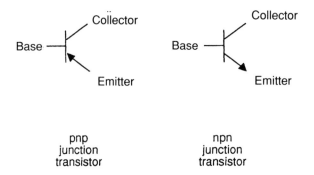

Figure 24.8 Circuit symbols for bipolar transistors

higher power applications the transistor is usually designed to be mounted on a metal heatsink, and power ratings from 1 to 200 W are available.

Field effect transistors

An alternative to the bipolar transistor is the field effect type. These have electrodes known as the source, gate and drain. The voltage between the gate and source electrodes controls the current flowing from drain to source. These transistors come in n channel and p channel varieties which correspond to the npn and pnp types in the bipolar range. A variation on the basic junction type FET is the MOSFET, where the GAT electrode is actually isolated from the n channel, which runs between source and drain. This gives the MOSFET an extremely high gate input impedance.

Field effect transistors have similar characteristics to thermionic valves and it is fairly easy to adapt valve-type circuits to operate with FETs. For high-frequency work the dual-gate MOSFET transistor can be useful, since it has characteristics similar to those of a tetrode valve, allowing the construction of stable RF amplifier stages. An example of this type of device is the 2SK88 n channel dual-gate MOSFET, which is widely used in amateur radio equipment.

MOSFETS are available as high-power devices with ratings up to 100 W. These are available in both p and n channel types, including matched p and n pairs which are popular for audio amplifier output stages. The VN66AF is a high-frequency type with a dissipation of 12.5 W and f_T of 600 MHz which can be useful for amateur transmitter projects.

Figure 24.9 shows the circuit symbols used for the various types of FET.

Thyristors and triacs

Thyristors or silicon-controlled rectifiers (SCRs) are solid-state devices which act like a rectifier diode when turned on by a signal applied to an extra

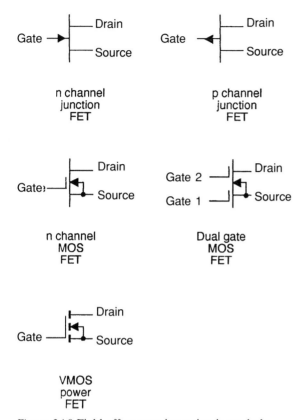

Figure 24.9 Field effect transistor circuit symbols

electrode called the gate. When the thyristor is not triggered it is effectively open circuit, and no current flows between anode and cathode. Once triggered, the thyristor will conduct whilst the anode remains positive relative to the cathode.

Thyristors can be used as solid-state relay switches and are widely employed for lighting control, motor speed control, and power supplies. When the thyristor is operating in an AC circuit it can be used to control the average power supplied to the load by altering the trigger point relative to the start of the positive half-cycle of the AC power waveform. As the trigger point gets later in the cycle, the pulse of current through the thyristor gets shorter and the power to the load falls. Thyristors are available with voltage ratings from 400 to 1000 V and current ratings from 1 to 100 A.

The triac is a variant of the thyristor which effectively contains two complementary thyristors connected in parallel. When triggered by a signal

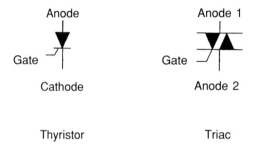

Thyristor   Triac

Figure 24.10 Thyristor and triac circuit symbols

on its gate electrode the triac will conduct on both positive and negative half-cycles of the AC power supply. Triacs are widely used for motor and lighting control and permit a wider range of power control, since they work on both half-cycles of the supply. Triacs are available in a similar range of voltage and power ratings to thyristors.

Figure 24.10 shows the circuit symbols used for thyristors and triacs.

Copper wire table

SWG number	Diameter (in)	Turns per in	Turns per cm	Ohms per 10 ft
16	0.064	15.0	5.9	0.025
18	0.048	19.8	7.8	0.044
20	0.036	26.1	10.2	0.079
22	0.028	33.3	13.1	0.13
24	0.022	42.1	16.6	0.21
26	0.018	50.6	19.9	0.32
28	0.0148	61.4	24.2	0.47
30	0.0124	73.3	28.9	0.66
32	0.0108	83.0	32.7	0.88
34	0.0092	98.0	38.6	1.21
36	0.0076	116.0	45.7	1.77
38	0.006	143.0	56.3	2.83
40	0.0048	180.0	70.9	4.40

Index